Natural Polymers for Biomedical Applications

Natural Polymers for Biomedical Applications

Wenguo Cui and Lei Xiang

WILEY-VCH

Authors

Dr. Wenguo Cui
Ruijin Hospital
Shanghai Jiao Tong University
School of Medicine
197 Ruijin 2nd Road
Shanghai 200025
China

Dr. Lei Xiang
Ruijin Hospital
Shanghai Jiao Tong University
School of Medicine
197 Ruijin 2nd Road
Shanghai 200025
China

Cover Image: © Andriy Onufriyenko/
Getty Images

Library of Congress Card No.: applied for

British Library Cataloguing-in-Publication Data A catalogue record for this book is available from the British Library.

Bibliographic information published by the Deutsche Nationalbibliothek The Deutsche Nationalbibliothek lists this publication in the Deutsche Nationalbibliografie; detailed bibliographic data are available on the Internet at <http://dnb.d-nb.de>.

Print ISBN: 978-3-527-35354-5

ePDF ISBN: 978-3-527-84557-6

ePub ISBN: 978-3-527-84559-0

oBook ISBN: 978-3-527-84558-3

Typesetting: Straive, Chennai, India
Printing and Binding: CPI Group (UK) Ltd,
Croydon, CR0 4YY

C9783527353545_260624

Contents

Graphical Abstract

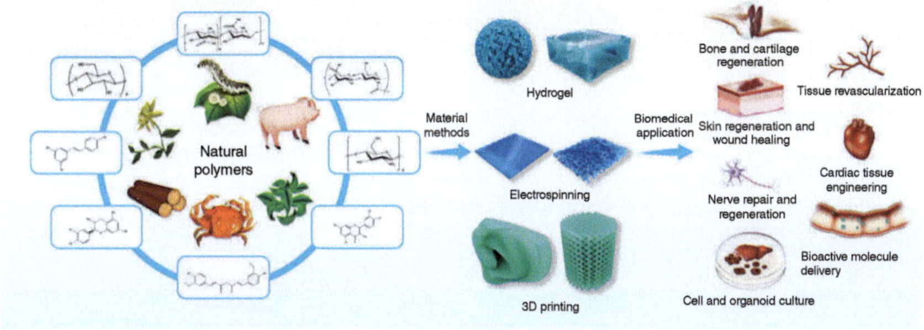

Natural polymer is a type of polymer compound formed based on photosynthesis or biochemical action that is widely found in animals, plants, and bacteria in nature, which has good biocompatibility, biodegradability, and low immunogenicity. This review summarizes the natural polymers basic structure and properties, focusing on the recent progress of natural polymer hydrogels, electrospun fibers, 3D bio-printing scaffolds, and polyphenol–metal complexes in the field of biomedicine.

Foreword

Source: Yuanjin Zhao (Chapter author).

Professor Yuanjin Zhao

With increasing research on natural polymers and the rapid development of biomanufacturing technology, various natural polymer-based biomaterials have been widely used to repair and regenerate bone, muscle, tendon, skin, and other tissues and organs. This book focuses on the current state of the manufacturing technologies and applications of natural polymer materials in the biomedical field.

The highlights of the book include a comprehensive analysis of different biomanufacturing methods for biomaterials related to polysaccharides, peptides, and polyphenol natural polymers, mainly covering recent developments in novel biomaterials such as hydrogels, electrospun fibers, metal-polyphenol nanoparticles, and 3D printing. In addition, the book provides a detailed overview of the advantages and disadvantages of each type of material for application in different scenarios and an entirely rationalized discussion of the future development and research directions of natural polymer-based biomaterials. Biofabrication technology effectively links clinical research and material development, helping to accelerate the clinical translation of novel natural polymer-based biomaterials and facilitating the development of targeted therapeutic strategies for different diseases or injuries. This book is a significant contribution to the advancement of translational medicine.

The book is organized in a precise, practical, rigorous, and, most importantly, up-to-date manner. It will provide the biomedical field with a reference on the preparation methods of natural polymer-based biomaterials and enhance the reader's understanding of natural polymers in biomedical applications.

In essence, this book is written in an informative manner, providing beneficial and relevant content for trained physicians and biomaterials researchers in the context of multidisciplinary cross-fertilization, which will help readers grasp the fundamentals of preparation of various biomaterials and thus deepen their understanding of their application in orthopedics. The authors believe that readers and researchers can gain inspiration and ideas for developing a new generation of natural polymer biomaterials through this book.

Yuanjin Zhao.

Yuanjin Zhao
Department of Rheumatology and Immunology, Nanjing Drum Tower Hospital, School of Pharmacy, Clinical College of Traditional Chinese and Western Medicine, Nanjing University of Chinese Medicine, Nanjing 210023, China
State Key Laboratory of Bioelectronics, School of Biological Science and Medical Engineering, Southeast University, Nanjing 210096, China
11 November 2023

Source: Kai Liu (Chapter author).

Professor Kai Liu

Over the past decades, inorganic polymers have been widely used in biomedical applications. However, such biomaterials have always suffered from a lack of bioactivity and poor biocompatibility, so there is an urgent need to innovate from the underlying material design logic. Natural polymer materials have attracted more and more attention from researchers due to their unique bioactivity and good biosafety, and their applications in biomedical fields are springing up.

This book systematically showcases popular and compelling natural polymer-based biomaterials and biomanufacturing technologies, explains in detail their current development and application potential in many clinical departments such as orthopedics, urology, dermatology, and endocrinology, and further analyzes the natural fit between natural polymer materials and various clinical diseases. One of the book's highlights is the coverage of the most innovative biomaterials manufacturing technologies, which emphasizes the safety of biomanufacturing, which we do not yet fully understand, and the solutions that need to be developed. The book also briefly explains why we should consider using these materials to treat diseases and efficiently repair and regenerate tissues and organs.

The editors and authors of this book hope to enrich the knowledge of frontline clinicians, readers in the biomedical field, and researchers in the field of bioengineering. This book can give many researchers and clinicians a detailed understanding of material science construction and ideas for developing novel biomaterials.

An excellent team consisting of the editors of this book, Prof. Wenguo Cui, Dr. Lei Xiang has made significant contributions to the field of biomedical engineering. The publication of this book will summarize the current applications of natural polymers in the biomedical field promptly and provide necessary references and inspiration to fellow researchers.

Kai Liu
Department of Chemistry, Tsinghua University, Beijing 100084, China
State Key Laboratory of Rare Earth Resource Utilization, Changchun Institute of Applied Chemistry, Chinese Academy of Sciences Changchun, Changchun 130022, China
20 November 2023

Preface

Source: Wenguo Cui (Book author)

Source: Lei Xiang (Book author).

The purpose of this book is to systematically review and comment on the latest research progress of various natural polymers in biomedical fields, covering the sources, structures, and preparation methods of different types of natural polymers, as well as their applications in drug delivery, cell culture, tissue engineering, etc., and providing detailed and informative valuable reference materials for the design and application of natural polymer materials. The research provides detailed, informative, and valuable references for the design and application of natural polymer materials. With the rapid development of the biomedical field in the past decades, natural polymers have attracted more and more attention from researchers because of their excellent biocompatibility, biodegradability, and low immunogenicity. Compared with synthetic polymers, natural polymers have unique molecular structures, abundant reactive groups, and biosignals, which can better mimic the microenvironment of cell growth in vivo and become ideal material platforms for tissue engineering and drug delivery systems. This book systematically describes the sources, structures, and properties of various natural polymers, including alginate, cellulose, hyaluronic acid, chitosan, gelatin, silk protein, starch, and dextran, and focuses on the recent advances in biomedical applications of these natural polymers in the past ten years, including the preparation, modification strategies, and drug delivery systems for natural polymer hydrogels, electrostatically spun fibers, 3D printed scaffolds, polyphenol–metal complexes. These include the preparation of natural polymer hydrogels, electrostatically spun fibers, 3D printed scaffolds, polyphenol–metal complexes, and their modification strategies and effects on drug delivery, cell culture, and tissue engineering.

Polysaccharide-based natural polymers have the ability to mimic extracellular matrix and provide support for cell adhesion and proliferation. Peptide such as collagen and gelatin contain special amino acid sequences, which can provide signals for cell adhesion. Polyphenolic compounds such as catechins and isoflavones have antioxidant pharmacological activities. By modulating the structure of these natural macromolecules and adding bioactive groups, researchers have prepared various biomaterials suitable for drug delivery, cell culture, and tissue regeneration, including injectable hydrogels, electrostatically spun fibrous membranes, and 3D-printed scaffolds. These biomaterial platforms have demonstrated excellent results in promoting cellular differentiation, controlled drug release, and repairing tissue damage in ex vivo and in vivo experiments.

We believe that the major contribution of this book lies in the comprehensive and systematic introduction of the structural features, preparation methods, and properties of various types of natural polymers, with a focus on summarizing the latest progress of their current biomedical frontiers, including hydrogels, electrostatic spinning, 3D bioprinting, and polyphenol–metal complexes, which can provide detailed and informative references for the design and application of natural polymers. We believe that with the accelerated cross-fertilization of biomedicine and materials science, natural polymers have great potential and development prospects in the biomedical field. The publication of this book will certainly promote the basic research and translational application of natural polymers and help to explore the potential of natural polymers in the era of smart medicine and precision medicine.

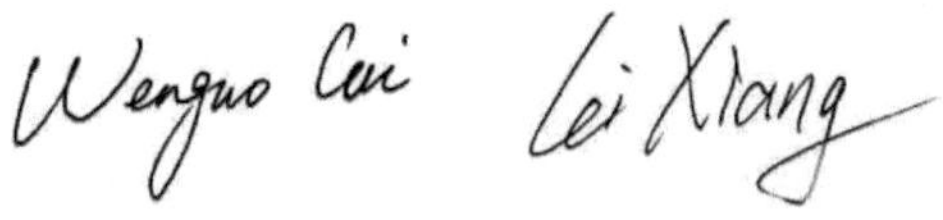

Wenguo Cui, Lei Xiang
Department of Orthopaedics, Shanghai Key Laboratory for Prevention and Treatment of Bone and Joint Diseases, Shanghai Institute of Traumatology and Orthopaedics, Ruijin Hospital, Shanghai Jiao Tong University School of Medicine, 197 Ruijin 2nd Road, Shanghai 200025, China
01 December 2023

Acknowledgments

This work was supported by grants from the National Key Research and Development Program of China (2020YFA0908200), National Natural Science Foundation of China (81930051), Shanghai Jiao Tong University "Medical and Research" Program (ZH2018ZDA04), and "The Project Supported by the Foundation of National Facility for Translational Medicine (Shanghai) (TMSK-2020-117)."

Section I

Historical Review of the Development of Natural Polymers

Natural polymer is a type of polymer compound formed based on photosynthesis or biochemical action that is widely found in animals, plants, and bacteria in nature. Compared with synthetic polymers, it has good biocompatibility, biodegradability, and low immunogenicity. Polysaccharides and polypeptides are often used as structural components because they can mimic the extracellular matrix environment and are rich in a variety of biologically active model sites, giving biological materials more biological activity and improving the biocompatibility of the materials. Polyphenols are often used as active ingredients in biomaterials due to their anti-oxidant, anti-free radical, and anti-cancer effects. There are various forms of biomaterials based on natural polymers. This book summarizes the basic structure and properties of natural polymers, focusing on the recent progress of natural polymer hydrogels, electrospun fibers, 3D bio-printing scaffolds, and polyphenol–metal complexes in the field of biomedicine.

Biomedicine covers many aspects such as drug and gene delivery, cell culture, and tissue engineering. With the continuous cross-fusion of biomedicine and materials science, many researchers are devoted to finding potential high-quality materials that can be used in the field of biomedicine, and more and more researchers have begun to use nature as a teacher to design and develop various types of bionic materials. In fact, bionics is not only limited to the simulation of the structure of some biological organs, but also includes the further development and utilization of natural materials that exist in nature [1, 2]. In the field of biomedicine, with the further development of various purification processes and continuous breakthroughs in the field of biochemistry, great changes have taken place in people's understanding of all kinds of natural active substances. Nature is just like a huge treasure, which contains many "pearls" with unique structure and activity that need to be discovered. Here, we mainly summarize the application of natural polymers in biomedical field in recent years.

Natural Polymers for Biomedical Applications, First Edition. Wenguo Cui and Lei Xiang.

Natural polymers have attracted more and more attention in biomedicine field in the past decades because of their advantages such as biodegradability, low toxicity, and low immunogenicity [3]. Natural polymer is a very broad concept, and we mainly introduce three types of natural polymers in this book: polysaccharides, peptides, and polyphenols. In this book, we will focus on the hot materials and technological forms in the biomedical field, including the cutting-edge applications of natural polymers in the fields of hydrogels, electrospinning, polyphenol–metal complexes, and 3D bio-printing.

At present, there are two main considerations when designing materials applied in tissue engineering field: one is whether the prepared material can meet the mechanical properties under normal physiological activities; the other is that on the basis of satisfying mechanical properties, whether it is conducive to cell proliferation and the microenvironment of tissue repair [4, 5]. The emergence of natural polymers development is how to design natural material which can well meet these two requirements. Natural polymers with different molecular weights and modification methods can be selected to achieve flexible control of their mechanical properties. Polysaccharide materials have signal regulation capabilities in cell membranes and cells [6, 7]. The special structure and protein sequence of peptides can mimic the extracellular matrix very well, creating external conditions for tissue regeneration [8, 9].

The design of materials in the field of drug and gene delivery generally considers the following points: First, protect the activity of the drug and deliver it to the target site. For example, oral drugs should avoid the destruction of drug activity by gastric acid and protect the intestines. The second is to achieve long-acting and sustained release of the drug and prolong the effective time of the drug [10]. The third is the low toxicity and functionalization of the drug carrier. The drug carrier needs to have low immunogenicity and synergistic or strengthening effect on the drug. The use of electrostatic forces between natural polymers and drugs to encapsulate or achieve sustained drug release by regulating the degradation rate has been widely used in the field of biomedicine [11]. In addition, there are many materials with environmental strain ability in natural polymers, such as temperature responsiveness, pH responsiveness, and enzyme responsiveness, and based on these properties, environmentally intelligent responsive drug carrier can achieve on-demand release of drugs. In addition, polyphenols have a certain pharmacological activity. Polyphenols, when combined with anticancer drugs such as doxorubicin, can have excellent synergistic effect, which can improve tumor cell killing rate, and the disadvantage of low bioavailability of polyphenols can be overcome by forming polyphenol–metal complex.

The purpose of this book is to summarize the recent application of natural polymer materials in cell culture, drug delivery, and tissue engineering, on the basis of introducing the source, structure, and basic properties of various natural polymers and hoping to provide a useful reference for researchers to design biomedical materials derived from natural polymers.

References

1 Zhao, D.W., Zhu, Y., Cheng, W.K. et al. (2021). Cellulose-based flexible functional materials for emerging intelligent electronics. *Adv. Mater.* 33: 2000619.

2 Giraldo, J.P., Landry, M.P., Faltermeier, S.M. et al. (2014). Plant nanobionics approach to augment photosynthesis and biochemical sensing (vol 13, pg 400, 2014). *Nat. Mater.* 13: 530.

3 Kadumudi, F.B., Hasany, M., Pierchala, M.K. et al. (2021). The manufacture of unbreakable bionics via multifunctional and self-healing silk-graphene hydrogels. *Adv. Mater.* 33: 2100047.

4 Matsuda, T., Kawakami, R., Namba, R. et al. (2019). Mechanoresponsive self-growing hydrogels inspired by muscle training. *Science* 363: 504–508.

5 Lin, W.F., Kluzek, M., Iuster, N. et al. (2020). Cartilage-inspired, lipid-based boundary-lubricated hydrogels. *Science* 370: 335–338.

6 Haas, K.T., Wightman, R., Meyerowitz, E.M., and Peaucelle, A. (2020). Pectin homogalacturonan nanofilament expansion drives morphogenesis in plant epidermal cells. *Science* 367: 1003–1007.

7 Tokatlian, T., Read, B.J., Jones, C.A. et al. (2019). Innate immune recognition of glycans targets HIV nanoparticle immunogens to germinal centers. *Science* 363: 649–654.

8 Matsumoto, A., Pasut, A., Matsumoto, M. et al. (2017). mTORC1 and muscle regeneration are regulated by the LINC00961-encoded SPAR polypeptide. *Nature* 541: 228–232.

9 Buszczak, M., Signer, R.A.J., and Morrison, S.J. (2014). Cellular differences in protein synthesis regulate tissue homeostasis. *Cell* 159: 242–251.

10 Miao, Y.B., Lin, Y.J., Chen, K.H. et al. (2021). Engineering nano- and microparticles as oral delivery vehicles to promote intestinal lymphatic drug transport. *Adv. Mater.* 33: 2104139.

11 Lin, F., Wang, Z., Xiang, L. et al. (2021). Charge-guided micro/nano-hydrogel microsphere for penetrating cartilage matrix. *Adv. Funct. Mater.* 31: 2107678.

Section II

Polysaccharides for Biomedical Application

Natural Polymers for Biomedical Applications, First Edition. Wenguo Cui and Lei Xiang.

1

Sources, Structures, and Properties of Alginate

Alginate, also known as alginic acid, is a kind of natural linear anionic polysaccharide widely used in the field of biomedicine. Alginate is mainly distributed in the cell wall of brown algae and extracted in the form of acidification and sodium salt. In recent years, some researchers have extracted alginate through microbial fermentation of *Pseudomonas aeruginosa* [1].

Alginate is a straight-chain polymer composed of 1-4-linked β-D-mannuronic acid (M) or α-L-guluronic acid (G), which is interspersed with regions containing alternating M–G sequence [2]. The chemical structure of alginate from different sources is also quite different. Alginate from algae has high content of Poly G and excellent antibacterial activity, while alginate from microorganism has high content of Poly M, which can induce monocytes to produce inflammatory mediators such as interleukin-1, interleukin-6, and tumor necrosis factor. Na^+ in alginate guluronic acid (Poly G) can be exchanged with divalent cations to form physically cross-linked hydrogels, in which the cross-linking model with Ca^{2+} is called "eggs-box model" (side-by-side Poly G can form a pore conducive to Ca^{2+} cross-linking, and each Poly G binds to the corresponding two Poly G in an orderly manner) [3]. During the cross-linking process, the alginate droplets containing the required protein will be extruded from the gel frame to form alginate microspheres. Due to the mild cross-linking conditions and excellent mechanical properties, alginate hydrogel has become a research hotspot in the field of cell encapsulation and tissue engineering. Owing to the free hydroxyl and carboxyl groups, alginate also has excellent bio-adhesion. In addition, the pH sensitivity of alginate (which shrinks under low pH conditions) also has a great prospect in targeted drug delivery.

At present, the modification of alginate is mainly based on the following two aspects: Firstly, the alginate materials with different properties were obtained by adjusting the content of Poly M and Poly G in alginate. Secondly, the active sites of alginate (carboxyl group, hydroxyl group, 1-4 glycosidic, internal glycolic bonds) were modified to improve the properties of the derived materials.

Natural Polymers for Biomedical Applications, First Edition. Wenguo Cui and Lei Xiang.

1.1 Alginate-Based Hydrogel for Biomedical Application

1.1.1 Drug and Cell Delivery

Systematic administration of antibiotics is the main cause of widespread drug resistance throughout the body, and the development of a local targeted administration system is an effective way to solve this clinical problem. In order to solve many side effects of intravenous application of antibiotics, Czuban et al. [4] prepared tetrazine-modified alginate hydrogel. Based on the principle of the inverse electron-demand Diels–Alder chemistry, vancomycin and daptomycin loaded with hydrogel can be released at the site of infection, and this hydrogel can repeatedly achieve drug loading and local release, significantly reducing adverse reactions caused by the use of antibiotics.

Autologous tumor cell vaccine is an individualized therapeutic strategy to activate tumor-specific immune response. However, it has limited efficacy in "cold" solid tumors that lack tumor-infiltrating T cells and are insensitive to immunotherapy. Ke et al. constructed a dendritic cell (DC)-activated hydrogel system using bifunctional fusion membrane nanoparticles (FM-NPs) composed of autologous tumor cell membranes and *Mycobacterium leprae* membrane extract to provide tumor antigenic signals and to interact with granulocyte-macrophage colony-stimulating factor (GM-CSF). Nanoparticles (NPs) composed of autologous tumor cell membranes and *Mycobacterium leucocephala* membrane extracts were used to provide tumor antigen signaling and were co-loaded with GM-CSF in an alginate hydrogel. Rapid release of GM-CSF recruited DCs; FM-NPs continuously activated the maturation of DCs and provided tumor antigens. The hydrogel system could increase the infiltration of effector memory T cells and activate "cold" tumors to exert significant anti-tumor effects. This study provides a feasible strategy to overcome the bottleneck of the efficacy of autologous tumor vaccines in "cold" tumors and points out a new direction to improve the clinical efficacy [5] (Figure 1.1).

The adenosinergic axis limits the effectiveness of current tumor immunotherapy by inhibiting the activity of effector T cells. How to effectively remodel the adenosinergic axis has become a key target to improve the effect of anti-tumor immunotherapy. Zhao et al. constructed an injectable hydrogel system based on alginic acid, and used the synergistic effects of adenosine deaminase, docetaxel, and benzotricarboxylic acid to realize the conversion from immunosuppressive adenosine to immuno-strengthening inosine to remodel the adenosinergic axis and exert anti-tumor effects. Docetaxel and benzotricarboxylic acid synergistically induced a large release of ATP, which triggered a strong immune response; adenosine deaminase catalyzed the conversion of adenosine to inosine, which further enhanced the immune effect; and ultimately achieved the reversal of the negative feedback from adenosine to positive feedback from inosine. The hydrogel strategy reshaped the adenosinergic axis through cascade amplification of ATP-mediated anti-tumor immune response, which provided a new idea and means to enhance the effect of tumor immunotherapy [6].

In the postoperative treatment of breast cancer, high local recurrence rates and potential wound infections pose significant risks to patient survival. To overcome

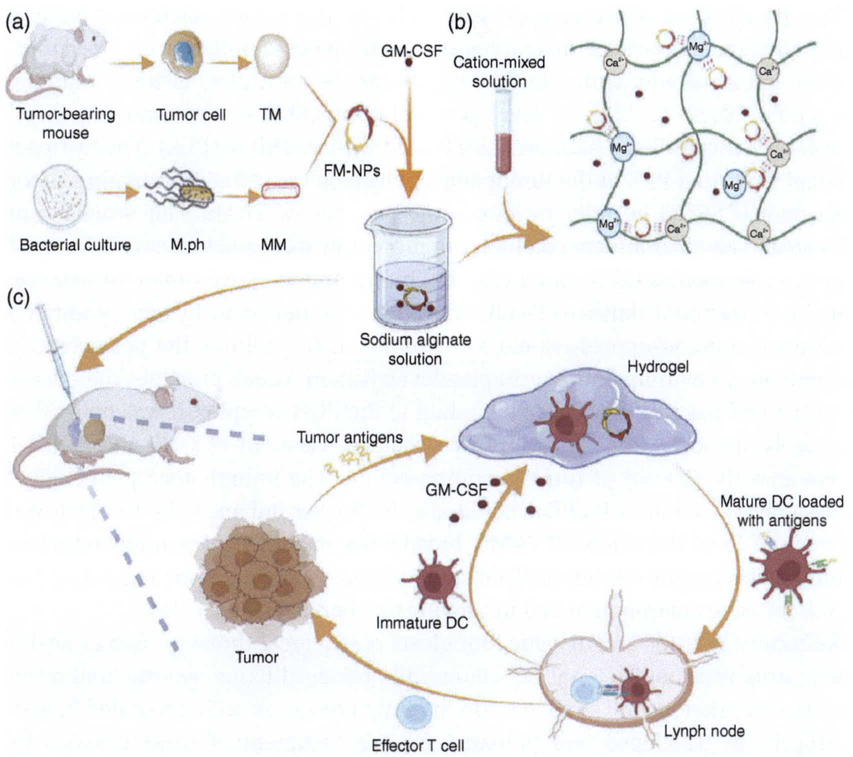

Figure 1.1 (a) Fusion membrane nanoparticles (FM-NPs) were prepared from autologous tumor cell membranes and *Bacillus* membrane extracts. (b) Sodium alginate solution was cross-linked with cationic solution to form hydrogels at room temperature. (c) FM-NPs and sodium alginate solution were used to form a hydrogel in vivo, which attracted dendritic cells and were activated by FM-NPs. Mature dendritic cells carrying tumor antigens stimulated the increase of effector memory T-cells, which exerted anti-tumor effects. Source: Ref. [5]/John Wiley & Sons.

these challenges, Wu et al. conducted a study on a nanocomposite dual-network (NDN) hydrogel. The hydrogel was constructed using polyethylene glycol acrylate (PEGDA) and alginate, embedded with 125i-labeled RGDY peptide-modified gold nanorods (125I-GNR-RGDY). This study formed hydrogels with a dual-network structure by near infrared (NIR) light-induced polymerization of PEGDA and endogenous Ca^{2+} cross-linking of alginate to construct a second network. This design enabled the hydrogel to exhibit stable photothermal effects and radiolabeling under NIR light irradiation. Photothermal therapy synergizes with brachytherapy by inhibiting DNA self-repair, promoting blood circulation, and improving the hypoxic microenvironment to enhance the therapeutic effect. This study provides a novel therapeutic approach by in situ injection of a precursor solution into the lumen of excised mouse cancerous breasts to form a rapidly gelatinizing hydrogel. By combining photothermal therapy and radiation therapy, this approach is expected to reduce the risk of local recurrence and decrease the likelihood of wound infection in postoperative breast cancer patients. This targeted therapeutic strategy offers new prospects for improving the outcome and survival of breast cancer patients [7].

The presence of immunosuppressive cells in the tumor microenvironment, especially tumor-associated macrophages (TAMs), poses a limitation on T-cell infiltration and activation, which in turn constrains the anticancer effect of immune checkpoint blockade. Li et al. developed a biocompatible alginate-based hydrogel that carries encapsulated nanoparticles loaded with pessitinib (PLX). The hydrogel gradually released PLX at the tumor site by blocking the colony-stimulating factor 1 receptor (CSF1R) in order to reduce the presence of TAMs. This strategy not only creates an environment conducive to promoting local and systemic delivery of anti-PD-1 antibodies (aPD-1), thereby inhibiting postoperative tumor recurrence, but also further contributes to T-cell infiltration of tumor tissue by reprogramming the tumor immunosuppressive microenvironment. In addition, the postoperative inflammatory environment triggers platelet activation, which promotes the release of aPD-1 and reactivates T cells by binding to the PD-1 receptor. It was noted that hydrogels can act as local reservoirs for sustained release of PLX-NP and P-aPD-1 to enhance the efficacy of tumor immunotherapy. The immunotherapeutic effect of systemic injection of P-aPD-1 could also be further enhanced by the hydrogel strategy of local depletion of TAMs, broadening the route of administration of immune checkpoint inhibitors. This study provides new ideas for regulating the tumor immune environment and improving the therapeutic effect [8].

Refractory keratitis and diabetic foot ulcers pose a great threat to human health due to drug-resistant bacterial infections and prolonged tissue hypoxia, and novel and effective therapeutic strategies are urgently needed. A self-oxygenated bilayer hydrogel was developed and prepared for the treatment of these diseases by Zhu et al. The inner hydrogel was composed of oxidized sodium alginate and carboxymethyl chitosan-containing photosensitizer PCN-224 and pH indicator bromothymol blue, while the outer hydrogel contained photosynthetic cyanobacteria. The inner hydrogel could sense the change of pH value to monitor the bacterial infection in real time and release PCN-224 in response to the infection for photodynamic bactericidal treatment; the cyanobacteria in the outer hydrogel continued to photosynthesize to produce oxygen to alleviate the hypoxic state of the tissues, enhance the effect of photodynamic therapy, and provide the necessary oxygen for the wound-healing process. In diabetic rat skin ulcer model and refractory keratitis animal model, the hydrogel could effectively sterilize, reduce inflammation, promote blood vessel regeneration and fibrous tissue formation, and significantly improve the therapeutic effect. The self-oxygenated bilayer hydrogel provides a novel strategy for the treatment of refractory ocular and skin diseases, and shows great application prospects and value [9].

Gan et al. used microfluidic electrospray technology to encapsulate mesenchymal stem-cell-derived exosomes in multilayered sodium alginate-gelatin (Gel) microcapsules for the targeted release of exosomes for protective delivery in the gastrointestinal tract and treatment of inflammatory bowel disease. In this study, inspired by the acid–base stability of Gel capsules, a novel multilayer microcapsule was prepared to encapsulate MSC-derived exosomes. The exosomes were first encapsulated in a core of sodium alginate gel microspheres using microfluidic electrospray technology and then coated with a Gel interlayer to protect them from degradation. The resistance of the microcapsules to gastric juices was enhanced by the use of a synthesized enteric polymer outer coating. The results showed that the prepared microcapsules could effectively protect the stability and bioactivity of exosomes against gastrointestinal digestion, and enable the release of exosomes at the site of intestinal injury to perform the biological functions of immunomodulation and damage repair. Therefore, the exosome-encapsulated microcapsules provide a new strategy of effective protection and targeted release for various oral cell therapies [10].

Zhang et al. designed an injectable hydrogel to simultaneously modulate T-cell exhaustion and MHC I expression for enhanced T-cell-based cancer immunotherapy. The hydrogel utilized sodium-oxidized alginate-modified tumor cell membrane vesicles (O-TMV) as the gelling agent and contained axitinib, 4-1BB antibody, and PF-06446846 nanoparticles. After the immune response was triggered by the O-TMV antigen, the hydrogel demonstrated superior immunotherapeutic effects through multiple mechanisms. 4-1BB antibody promoted T-cell mitochondrial biogenesis, axitinib reversed T-cell exhaustion, and PF-06446846 amplified MHC I expression to improve T-cell recognition of tumor cells. O-TMV@ABP hydrogel effectively inhibited tumor growth through strong immune responses and long-term memory immune response effectively inhibits tumor growth, metastasis, and recurrence. This innovative strategy provides a new concept for T-cell-based cancer immunotherapy and demonstrates the strong potential of the hydrogel platform [11].

Stem cell injection therapy has significant efficacy in the treatment of many diseases such as diabetes, but simple stem cell injection has the problems of immune rejection and graft removal. In order to solve these problems, Delcassian et al. [12] loaded COOH-modified iron oxide nanoparticles and living islet cells into alginate hydrogel. On the one hand, the immune rejection of alginate hydrogel is significantly reduced; on the other hand, iron oxide nanoparticles can make the inhibitor move directionally in the magnetic field, which solves the problem of recovery after failed transplantation (Figure 1.2).

An adhesive and adjustable methacrylic-acid-modified alginate saline gel has been developed by Hasani-Sadrabadi et al. [13], which was coated with gingival mesenchymal stem cells and hydroxyapatite particles were introduced to induce bone regeneration, which effectively repaired mouse craniofacial bone defects. Whitehead Jacklyn et al. [14] reported that hydroxyapatite nanoparticles adsorbed BMP-2 were added to mesenchymal stromal cell spheres and then embedded in Arginine–Glycine–Aspartic acid (RGD)-modified alginate hydrogels. They found that the viscoelastic dynamic mechanical properties of alginate hydrogels obtained by ionic cross-linking significantly enhanced the therapeutic potential of MSCs spheres in bone formation and repair. Hung et al. [15] have prepared a peptide-modified viscous hydrogel which can double stimulate MSCs. On the one hand, QK peptides are used to functionalize alginate to promote the secretion of angiogenic factors; on the other hand, RGD modification promotes cell adhesion and proliferation.

Hasturk et al. prepared enzyme-cross-linked alginate and alginate/gelatin composite microspheres by a simple and economical centrifugation method for microencapsulation protection of mammalian cells. The composite microspheres were more structurally stable under ionic conditions and had better mechanical properties than the ionically cross-linked alginate microspheres only. It was shown

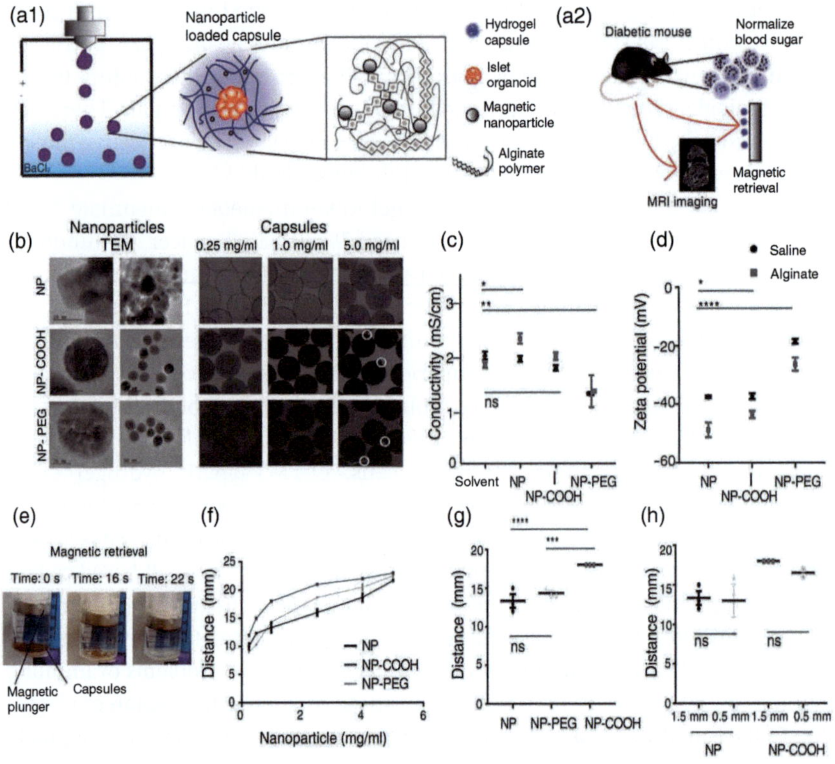

Figure 1.2 (a1) Schematic diagram of hydrogel vesicle formation containing nanoparticles; (a2) MRI and magnetic retrieval for diabetic transplantation. (b) TEM images of different nanoparticles and microscopic images of hydrogel capsules containing 0.25–5 mg/ml nanoparticles. (c,d) Conductivity and zeta potential of nanoparticles in saline or saline-sodium alginate. The conductivity and zeta potential of the nanoparticles were reduced compared to that of sodium alginate. (e–h) Magnetic recovery experiments were performed on hydrogels containing nanoparticles. As the concentration of nanoparticles increased, the traveling distance of the hydrogel increased; decreasing the capsule size decreased the traveling distance. Source: Ref. [12]/John Wiley & Sons.

that the human mesenchymal stem cells and neural progenitor cells encapsulated in the composite microspheres were effectively protected against various environmental factors, including extracellular toxins, acidosis, apoptotic factors, ultraviolet radiation, hypoxic conditions, and mechanical stresses. The microencapsulated cells maintained high viability, proliferation, and directed differentiation after extrusion through a 27-gauge needle. This demonstrated that the novel microencapsulation strategy has promising applications in cell injection delivery and three-dimensional bioprinting. Overall, the double-cross-linked composite microspheres provide a new, simple, and effective method for microenvironmental regulation and protection of mammalian cells [16].

Myocardial infarction (MI) is a major cause of sudden cardiac death, and platelet-rich fibrin is beneficial for restoring vascular regeneration in the infarcted area due to its richness in growth factors. Based on its biocompatibility and cost-effectiveness, alginate hydrogel is an advantageous way to achieve targeted delivery of platelet-rich fibrin. Qian et al. achieved infiltration of M2 macrophages in the infarcted area through the construction of alginate hydrogel, which is beneficial for improving the degree of myocardial fibrosis, and this hydrogel can also provide strong mechanical support for the ventricular wall and improve cardiac function [17].

Localized stem cell delivery to the human locomotor system and major weight-bearing tissues requires high structural strength of the delivery platform. Panebianco et al. used degradable alginate microspheres to microencapsulate mesenchymal cells to protect the cells and maintain cell viability and phenotype upon release. The composite of cell-carrying microspheres with high-modulus cross-linked fibrin gels balanced biomechanical properties and cell biological activity. The composites showed better cell survival and matrix synthesis than fibrin gels alone. The biomechanical stability of the composites and their ability to promote extracellular matrix synthesis were verified in large animal in vitro experiments. The composites significantly improved biomechanical function and biological repair compared with discectomy alone [18].

Diabetic foot ulcer (DFU) is a serious complication of diabetes mellitus, and the local hypertonicity makes the wound prolonged, Theocharidis et al. thoroughly investigated the potential application of alginate dressings for the local delivery of macrophages and their secretory products for the treatment of DFU. By preparing alginate dressings with a microporous structure, they were able to achieve a uniform loading of primary macrophages and realized the loading of macrophages with different polarization states (M0, M1, M2a, M2c) onto the dressing and their migration into the wound. The experimental results demonstrated that the treatment of each macrophage subtype promoted DFU healing in db/db mice [19].

1.1.2 Cell and Organoid Culture

Different hydrogel micropatterns can be formed by adjusting the cross-linking density of various parts of the hydrogel. Jeon et al. [20] used chemical and optical double-cross-linking methods to give oxidized, methacrylated alginate and 8-arm poly amine hydrogel micropatterns of different sizes, and found that the size of these micropatterns can significantly affect the proliferation and differentiation behavior of cells. The larger the micropattern (25–100 μm), the more obvious the proliferation and osteoblast differentiation of human adipose-derived stem cells. In addition, Gonzalez-Pujana et al. [21] further studied the effect of hydrogel on controlling the gene expression and secretory behavior of human marrow mesenchymal stem cells (HMSCs). They prepared click functionalized sodium alginate and fibrous collagen composite hydrogels and loaded with interferon-γ and heparin-coated microspheres. In the hydrogel microenvironment, bone marrow mesenchymal stem cells increased the expression of key genes such as indoleamine 2-dioxygenase-1 and galactose lectin-9, and promoted the secretion of license response factor Gal-9. And HMSCs cultured in hydrogel can inhibit activated human T cells, which proves that this hydrogel can enhance the immunomodulatory properties of HMSCs. Liu et al. [22] developed a hydrogel microfluidic system based on the interfacial complexation of inversely charged sodium alginate and chitosan, and used this capsule to culture human induced pluripotent stem cells (HiPSCs), and successfully formed islet organisms (containing islet-specific α and β-like cells, highly expressing pancreatic hormone-specific genes and proteins). It reveals the great potential of alginate hydrogel in the preparation of artificial organs (Figure 1.3).

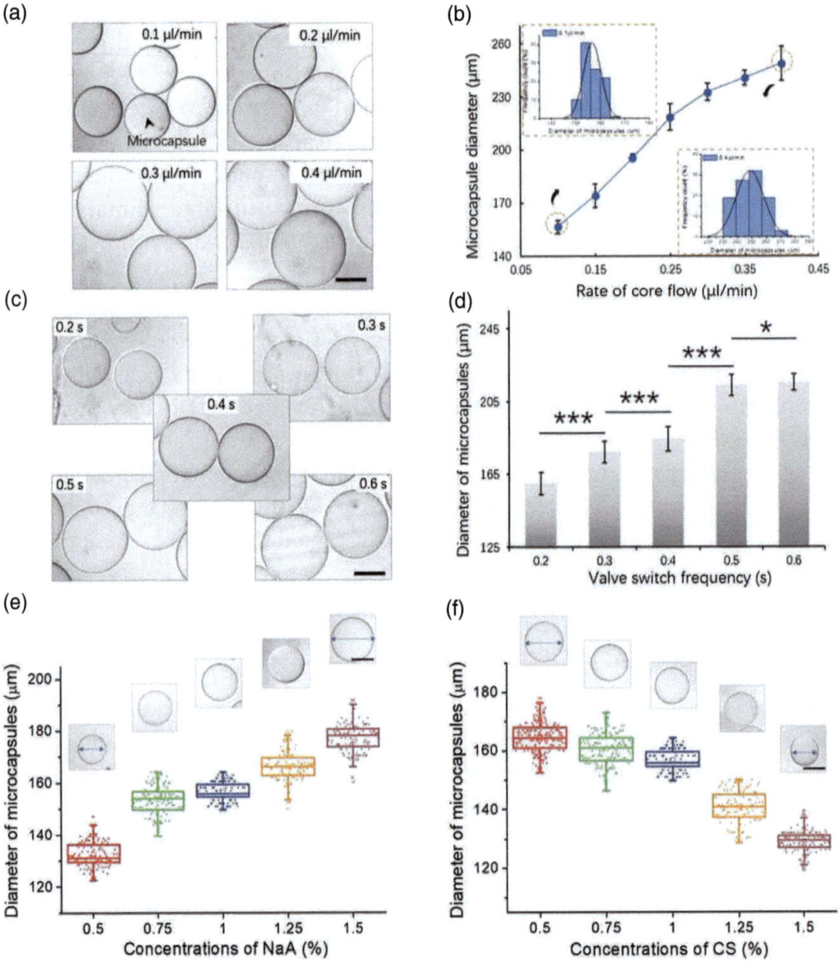

Figure 1.3 (a) Images of hydrogel vesicles generated at different core flow rates. The hydrogel vesicle size decreases with increasing core flow velocity. (b) At a fixed boundary flow rate, the hydrogel vesicle size is negatively correlated with the core flow rate. Increasing the core flow rate can reduce the size distribution of hydrogel vesicles. (c) Hydrogel vesicle images generated at different valve switching frequencies. The hydrogel vesicle size decreases with increasing frequency. (d,e) Hydrogel vesicles generated with different concentrations of sodium alginate. The size of the hydrogel vesicle decreases as the concentration of sodium alginate increases. (f) Hydrogel vesicles generated by using different concentrations of chitosan. The size of the hydrogel vesicles decreases as the concentration of chitosan increases. Source: Ref. [22]/John Wiley & Sons/CC BY 4.0.

1.1.3 Tissue Regeneration

The repair of cartilage-bone defects has always been a challenging task. Liao et al. [23] used alginate and calcium gluconate to cross-linking the upper hydrogel (composed of CSMA and NIPAM) and the lower hydrogel (composed of PECDA, AAm, and PEGDA) to form bidirectional hydrogels. The problem that the junction tissue is difficult to repair is perfectly solved by a two-way hydrogel design (the lower layer promotes bone regeneration and the upper layer promotes cartilage regeneration). Öztürk et al. [24] reported sulfated alginate hydrogel can induce the rapid proliferation of chondrocytes through gene-mediated FGF receptors and downstream signals, which opens up a new idea for cartilage repair. When the cartilage is so damaged that it is difficult to repair by its own tissue cells, the defect can only be repaired by artificial cartilage implantation (Figure 1.4).

Liao et al. [25] used the inter-transmission network structure formed by sodium alginate, fibrin, and polyacrylamide to prepare a hydrogel similar to that of natural cartilage with friction coefficient, and the implanted hydrogel is beneficial to

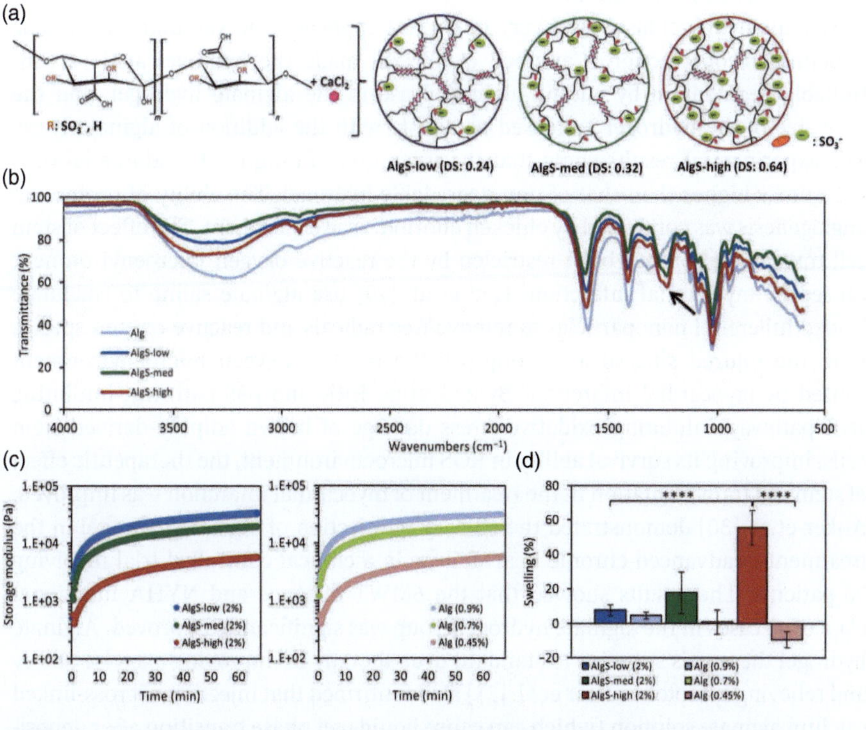

Figure 1.4 (a) Schematic of the chemical structure of sulfated sodium alginate. The degree of sulfation increases and cross-links with calcium ions to form an alginate sulfate hydrogel. (b) Infrared spectra of sodium alginate with different degrees of sulfation. The S=O characteristic peaks are enhanced by increasing the degree of sulfation. (c) Rheological profiles of alginate sulfate and sodium alginate hydrogels. The storage modulus of the hydrogels increases with the increase of the degree of sulfation. (d) Mass swelling percentage of the two hydrogels. The swelling percentage of alginate sulfate hydrogel is lower than that of sodium alginate hydrogel. Source: Ref. [24]/John Wiley & Sons.

cell infiltration and angiogenesis. It has great potential in artificial cartilage transplantation.

Dental tissue stem cells not only have great potential in promoting bone repair, but also play a great role in tendon regeneration. Moshaverinia et al. [26] prepared TGF β 3-RGD-coupled alginate microspheres as carriers to construct a co-transport system of periodontal ligament stem cells and gingival mesenchymal stem cells. A large number of regenerated tendon tissue deposits were found by injecting modified alginate microspheres into immunocompromised mice. Mredha et al. [27] study in-depth of alginate hydrogel and further confirmed its role in the preparation of artificial tendons. By adding calcium ions to the alginate solution to form a physical gel, and while drying the hydrogel, the tractive force in the length direction is applied at both ends of the hydrogel, so that the length of the hydrogel remains unchanged while the width and thickness of the hydrogel are reduced during the drying process. In this way, the hydrogel forms a hierarchical fiber structure, and further drying will induce the nanofibers to gather and form thicker fibers. Due to the stable supramolecular interaction between polymers, the re-swollen gel maintains the same structure as the natural tendon.

Alginate hydrogel also has a very important application in vascular and cardiac structural reconstruction. Campbell et al. [28] made the hydrogel achieve controllable degradation by adding alginate lyase to the alginate hydrogel, and the pore size of the hydrogel increased obviously with the addition of alginate lyase. The experimental results show that the migration of outgrowth endothelial cells is 10 times higher than that of non-degradable hydrogel. The ability of promoting angiogenesis was confirmed by chicken chorionic sac experiment. The effect of stem cell transplantation has been restricted by the reactive oxygen microenvironment caused by myocardial infarction. Hao et al. [29] use alginate saline to coagulate loaded fullerenol nanoparticles to remove free radicals and reactive oxygen species from the injured site, so as to improve the reactive oxygen microenvironment caused by myocardial infarction. By activating ERK and p38 pathway, inhibiting JNK pathway, inhibiting oxidative stress damage of brown adipose-derived stem cells, improving its survival ability in ROS microenvironment, the therapeutic effect of stem cell transplantation in the treatment of myocardial infarction was improved. Anker et al. [30] demonstrated the efficacy of injection of alginate hydrogel in the treatment of advanced chronic heart failure in a clinical controlled trial involving 78 patients. The results showed that the 6MWT distance and NYHA functional class of 40 cases in the alginate hydrogel group was significantly improved. Alginate hydrogel therapy is superior to standard drug therapy in improving exercise ability and relieving symptoms. Leor et al. [31] also confirmed that injection of cross-linked calcium alginate solution (which can cause liquid–gel phase transition after deposition in the infarcted myocardium) four days after infarction can effectively prevent ventricular enlargement and promote myocardial remodeling.

Stem-cell-based tissue engineering offers new therapeutic avenues for the treatment of various diseases and tissue regeneration; however, the infusion of exogenous stem cells still has many problems such as limited cell sources and potential rejection. Therefore, how to maximize the mobilization of residual stem cells around the defect site may be the key to solving the problem of tissue regeneration. He et al. utilized the acoustic response of alginate hydrogel to load BMP-2, which has the ability to recruit stem cells, and used pulsed ultrasound to stimulate the intrinsic resonance of the hydrogel to achieve rapid degradation of the hydrogel and on-demand recruitment of endogenous stem cells [32].

Conventional implantable block hydrogels for cavernous injuries cannot meet the needs of repairing irregular lesions, so injectable hydrogels with shear-thinning properties have a more suitable application environment. Zheng et al. prepared an injectable hydrogel system based on bioglass, alginate, and filipin proteins. This hydrogel can achieve responsive degradation according to the ionic concentration of the biological environment, as well as anti-inflammatory and pro-regenerative effects through the MAPK signaling pathway [33].

Wang et al. synthesized conductive hydrogels using silver nanowires (AgNW) and methacrylic acid alginate (MAA) to efficiently promote wound healing by using electrical stimulation. Meanwhile, the incorporation of silver nanowires greatly improved the structural steeliness of the soft electronic material. By applying localized electrical stimulation to the wound area, rapid wound closure could be achieved within seven days, and the expression of growth factors in NIH-3T3 cells could be promoted, as well as the proliferation and migration of NIH-3T3 cells [34].

A novel dual-network sodium alginate-platelet-rich plasma hydrogel was designed and prepared for the promotion of wound healing by Wang et al. The hydrogel was prepared by a simple one-step thrombin activation process and had a three-dimensional network structure. The hydrogel can release growth factors that promote cell proliferation and vascular regeneration. Topical application of the hydrogel to rat wounds significantly increased the rate of wound closure. The dual-network hydrogel provides a simple and effective strategy to overcome the shortcomings of traditional wound dressings [35].

Tendons are composed of soft collagen, whose anisotropic structure and tendon-osteointegration allow for a strong attachment to bone. To mimic this property, Choi et al. developed a tough triple-network (TN) hydrogel realized by a dual network of imidazole-containing polyaspartic amide and alginate-polyacrylamide. The hydrogel exhibited high tensile modulus and strength on the bone surface while maintaining excellent bone adhesion without the need for chemical treatment of the bone surface. By introducing a third polymer, a bone–ligament–bone structure resembling natural ligaments was also successfully realized [36].

The treatment of skin wounds in joints and moving parts has always been a difficult problem. Due to frequent movement and flexion, wound closure using conventional sutures and skin adhesives not only causes additional trauma, but also anesthetic side effects and severe scarring. Liu et al. designed and prepared a novel hydrogel composed of THMA, PEGDA, and sodium alginate. This hydrogel utilized hydrogen bonding and chemical cross-linking to form an interpenetrating network structure, which gave it ultra-high elongation (>700%) and surface adhesion (7.5 kPa) for long-lasting apposition of high-frequency moving parts, and its transparent gel appearance facilitated the observation of wound recovery [37].

To address the problem that hydrogels are not directly absorbed and utilized by wounds, Yao et al. developed a novel histidine-based healing hydrogel. Histidine is a natural dietary essential amino acid that is highly beneficial for tissue formation. Through dynamic coordination and hydrogen bonding, histidine was cross-linked with zinc ions (Zn^{2+}) and sodium alginate (SA) to form a histidine–SA–Zn^{2+} (HSZH) hydrogel. The hydrogel exhibited good injectability, adhesion, biocompatibility, and antimicrobial properties. The HSZH hydrogel was cross-linked with double dynamic bonds, which accelerated the migration and angiogenesis of skin-associated cells in vitro. In vivo experiments demonstrated that the hydrogel significantly facilitated the healing of infected diabetic wounds, taking only about 13 days to fully repair the wounds, compared to the healing process in the control group, which took about 27 days. This study provides new ideas for the design of wound dressing materials, in which weakly cross-linked materials based on tissue-friendly micromolecules are more effective in promoting wound healing compared to highly cross-linked materials based on long-chain polymers [38] (Figure 1.5).

1.1.4 Other Applications

In addition, alginate hydrogel still has many applications in cancer targeted therapy, in vitro imaging, skin repair, and hemostasis. Wang et al. [39], through preparing sodium alginate-calcium hydrogel and loading platinum nanoparticle in its matrix, can degrade the hydrogel on demand and release chemotherapeutic drugs through repeated photothermal therapy, which can improve the therapeutic effect of cancer. Patrick et al. [40] made use of the natural metal chelation characteristics of alginate to creatively cross-link alginate hydrogel with radioactive metal cations In^{3+} and Zr^{4+}, realized the detection of alginate hydrogel in vitro, and demonstrated its in vivo nuclear imaging of longitudinal retention and clearance of alginate hydrogel in oral and nasal administration, stem cell transplantation, and heart tissue engineering. By preparing bio-glass and sodium alginate composite hydrogel, Zhu et al. [41] found that the hydrogel can polarize macrophages to M2 phenotype and up-regulate the expression of anti-inflammatory genes. In addition, it can also recruit fibroblasts and endothelial cells to accelerate the repair process of full-thickness skin. Ren et al. [42] reported coating alginate calcium chloride hydrogel on the needle can significantly reduce the incidence of puncture bleeding, and confirmed its function of in situ hemostasis after tissue puncture in vein, kidney, and liver puncture experiments, which provides a feasible scheme for improving the safety of clinical tissue biopsy.

A study by Jons et al. evaluated the rheological characteristics of calcium alginate and polymer–nanoparticle gels and correlated them with their ability to form drug reservoirs after subcutaneous administration in mice. The rheology of the tested gels included stiffness, viscoelasticity, yield stress, and creep behavior. By combining the rheological characteristics of the gels with the effects of in vivo administration, it was found that yield stress predicted the initial formation of reservoirs and creep predicted the persistence of reservoirs, with yield stresses >25 Pa resulting in the formation of solid reservoirs. This provides a predictive reference for the design

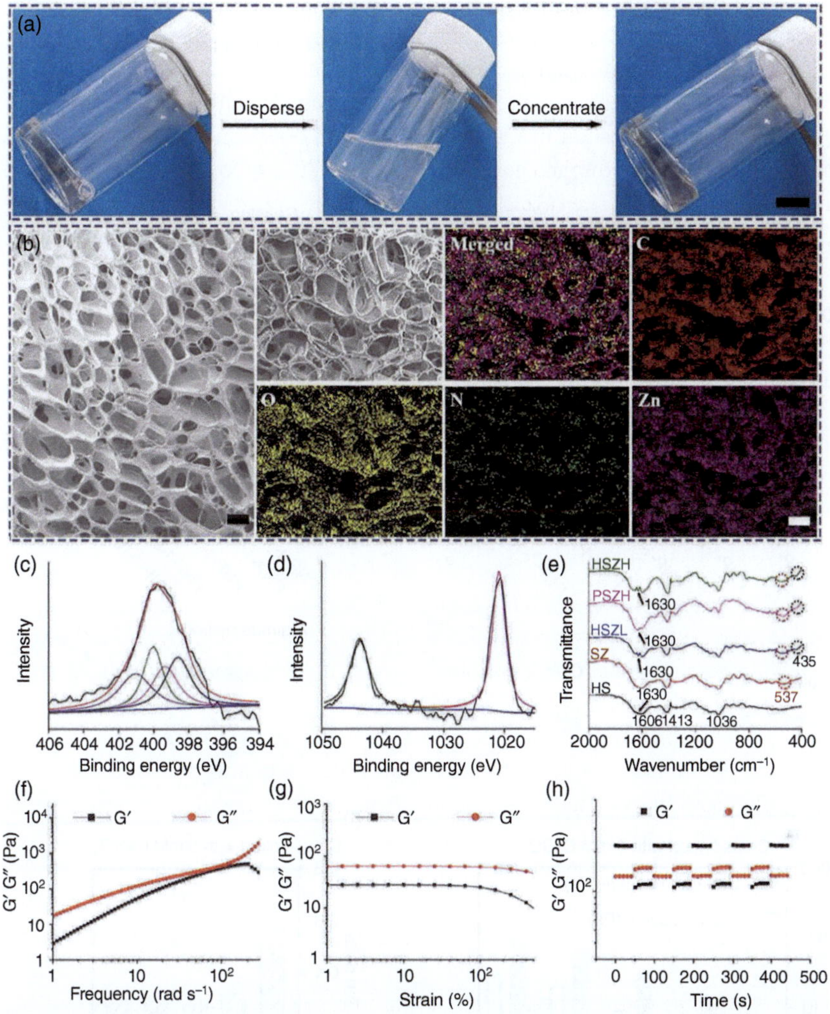

Figure 1.5 (a) The hydrogel is reversible and can be formed and dissolved repeatedly. (b) Scanning electron microscopy images of HSZH hydrogel showed a porous network structure, and elemental analysis confirmed the inclusion of Zn. (c) X-ray photoelectron spectroscopy analysis showed that the HSZH hydrogel contained nitrogen. (d) X-ray photoelectron spectroscopy analysis showed that the HSZH hydrogel contained Zn. (e) Infrared spectroscopy analyzed the characteristic peaks of different hydrogels. (f–h) Rheological tests show that HSZH hydrogels have frequency-dependent, strain-dependent, and step-strain oscillatory shear rheological properties. Source: Ref. [38]/John Wiley & Sons.

of hydrogel systems for sustained controlled release. Prospect: This study reveals the relationship between rheological characteristics and in vivo drug reservoir formation and durability, and provides guidance for the future design and optimization of the rheology of hydrogel controlled-release systems, which is expected to promote the advancement of related technologies and ultimately benefit patients [43] (Figure 1.6).

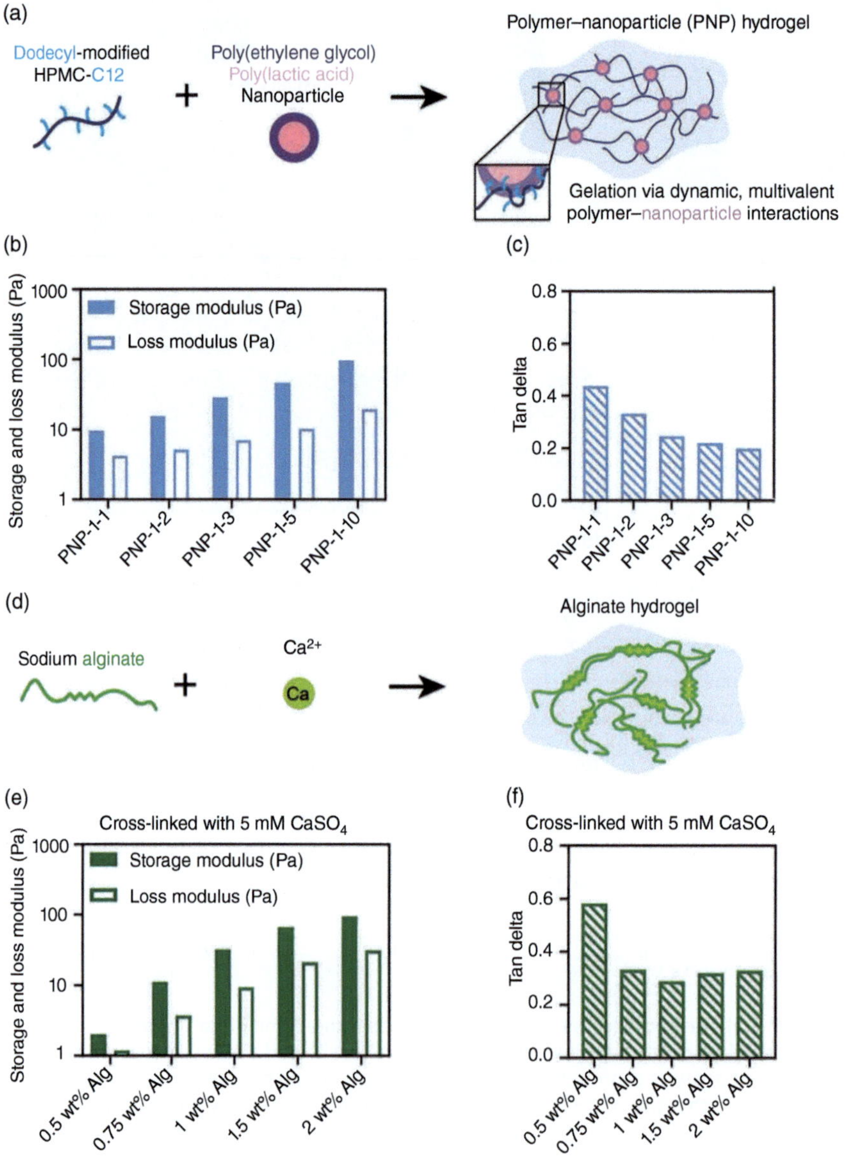

Figure 1.6 (a) A hydrogel preparation method formed from polymers and nanoparticles by non-covalent cross-linking. (b) Energy storage modulus and loss modulus of PNP hydrogel formulation. Increasing the nanoparticle content increases the energy storage modulus. (c) tanδ of PNP hydrogel formulation. Increasing nanoparticle content can decrease tanδ. (d) A hydrogel preparation method formed by cross-linking sodium alginate and calcium ions. (e) Energy storage modulus and loss modulus of sodium alginate hydrogel formulation. Increasing the calcium ion concentration increases the energy storage modulus. (f) tanδ of sodium alginate hydrogel formulation. Increasing calcium ion concentration can decrease tanδ [43].

Ji et al. reported a simple reconstruction process to prepare alginate hydrogels with ultra-strong, ultra-hard, and conductive properties, which can be widely used in artificial biological tissues, flexible electronic devices, and conductive membranes. Through anisotropic densification of the pregel and subsequent ionic rehydration ion cross-linking, the reconstructed hydrogels exhibit exceptional tensile strength (8–57 MPa) and elastic modulus (94–1290 MPa) depending on the type of cross-linked ions. Such hydrogels are able to accommodate sufficient cations (e.g. Li^+) without affecting their mechanical properties and exhibit good ionic conductivity, which is suitable for the preparation of gel electrolyte membranes. In addition, we demonstrate the incorporation of conducting polymers into hydrogel matrices to prepare ionic/conducting hydrogels with outstanding mechanical properties. Through simple surface de-crossing and re-crossing, such hydrogels have strong interfacial adhesion. In conclusion, we have developed mechanically enhanced hydrogels by a simple reconstruction method that forms tightly connected polymer networks. This simple process is suitable for large-scale production and can be expected to be used in practical manufacturing applications [44].

Microwave ablation (MWA) is a local tumor treatment strategy, but is often challenged by tumor recurrence. Therefore, the development of adjuvant biomaterials to enhance the effectiveness of MWA is relevant. The results showed that alginate-immobilized Ca^{2+} formed hydrogels under microwave exposure, exhibiting efficient heating and a restricted heating zone. High extracellular Ca^{2+} concentrations synergized with microwave subthermal therapy to induce immunogenic cell death by interfering with intracellular Ca^{2+} homeostasis. Thus, Ca^{2+}-remaining alginate hydrogel combined with MWA can effectively ablate different tumors in mice and rabbits while reducing surgical trauma. This treatment also triggered an anti-tumor immune response, especially when combined with interferon gene pathway activators, which inhibited the growth of untreated distant and recurrent tumors. This study highlights the potential of metal alginate hydrogels as biomaterials for microwave therapy and immunostimulation, with promising clinical applications [45].

1.2 Alginate-Based Electrospinning for Biomedical Application

1.2.1 Drug Delivery

Electrospinning can well simulate the structure of extracellular matrix, coupled with its unique porous structure, so it is widely used in the field of tissue engineering and drug delivery. However, alginate is easy to gelate; low content and impurities restrict its application in electrospinning technology. At present, most researchers improve its electrospinning properties by adding alginate into polymer solution. In addition, Asadi-Korayem et al. [46] further studied the effect of intermolecular hydrogen bonding on the properties of alginate electrospinning. According to the hard and soft acids and bases theory, they used Li^+ instead of Na^+, to further strengthen the interaction between ions and reduce the hydrogen bond density, which significantly improved the electrospinning properties of alginate. Fujita et al. [47] successfully prepared alginate nanofibers by wrapping sodium alginate in a nanofiber shell, then gelating sodium alginate and removing the shell core electrospinning method, and immobilized fibronectin on the resulting alginate fibers to control the proliferation of cells along the fiber direction.

Kaassis et al. [48] blended polyoxymethylene, sodium alginate, and ibuprofen in the preparation of electrospun fibers. They found that ibuprofen microcrystals were present on the electrospun fibers prepared by this method, and the experimental drugs were pulsed in different pH environments. They can control the release amount and release interval of the fiber at different stages by adjusting the loading of sodium alginate and ibuprofen. De Silva et al. [49] loaded cefalexin onto cemented carbide nanotube to enhance the mechanical properties of alginate-based nanofiber scaffolds. The addition of cemented carbide nanotubes extended the drug release time from 24 hours to 7 days, and showed strong broad-spectrum antibacterial properties against Gram-positive and Gram-negative bacteria.

Dodero et al. [50] used electrospinning technology to prepare polycaprolactone (PCL)- and sodium alginate-embedded nano-ZnO bilayer fiber films. The outer layer of polycaprolactone has good hydrophobicity, while the inner layer of sodium alginate can promote tissue regeneration and remove wound exudates. In addition, they have verified the drug-carrying capacity of the fiber membrane, and the release form of the drug can be controlled by adjusting the concentration of the loaded drug. Mulholland et al. [51] used siRNA targeting FK506-binding protein-like proteins to promote angiogenesis. Chitosan-sodium alginate double-layer wound patches were prepared by electrospinning. They found that the blood vessel density in the treatment group increased by 326% compared with the control group, confirming the great role of siRNA targeting FK506-binding protein-like proteins in wound healing. Tang et al. [52] introduced honey into sodium alginate/polyvinyl alcohol (PVA) electrospun nanofiber membrane to give the fiber membrane stronger antioxidant activity and antibacterial activity. Similarly, Hajiali et al. [53] prepared wound dressings with both antibacterial and anti-inflammatory effects by adding essential oils. On the basis of preparing alginate electrospun films containing ZnO nanoparticles, Dodero et al. [54] further explored the effects of different cross-linking agents on the properties of fiber membranes. They found that compared with the most commonly used Ca^{2+}, Sr^{2+} has a higher affinity for alginate, can improve and accelerate tissue regeneration, and the fiber membrane cross-linked by Sr^{2+} is closer to human skin in mechanical properties. Electromagnetic action is also an emerging way to achieve targeted drug delivery, and Chen et al. [55] prepared magnetic induction fiber membrane by chelating alginate electrospun fiber with Fe^{2+}/Fe^{3+}, which showed targeted killing effect on tumor cells in alternating magnetic field.

Alloisio et al. successfully synthesized alginate chain-stabilized silver nanoparticles (Alg@AgNPs) in situ in polysaccharide solutions using wet chemistry as a powerful alternative to antimicrobial materials. This nanocomposite material combines the efficient and broad-spectrum biocidal properties of silver nanoparticles with the biocompatibility and environmental friendliness of natural polysaccharide components. Nonwoven films with uniform nanostructures with fiber diameters between 100 and 150 nm, influenced by the size of the embedded metal nanoparticles (between 20 and 35 nm), were prepared by electrostatic spinning technique. Preliminary tests showed that these nanocomposite films exhibited a biocidal effect against Gram-negative *Escherichia coli* (*E. coli*) that could occur within one day and was observed even when the content of AgNPs in the polysaccharide fibers was well below nanomolar levels [56].

Among numerous tissue engineering applications, electrospun fibers are considered as a material with good potential to be used as wound dressings, which help to promote the healing process and maintain the regeneration of damaged skin. However, conventional electrospun fibers have some drawbacks in wound repair, such as the lack of antibacterial, anti-inflammatory, and angiogenesis-promoting properties. Lashkari et al. prepared bilayer scaffolds consisting of PCL/Gel nanofibers and collagen alginate (Col/Alg) hydrogel and implanted adipose-derived stem cells (ADSCs) into the nanofibers and the bilayer scaffold. The double-layer scaffold with implanted ADSCs and nanofibers was used to dress the whole level of trauma on the back of rats. Histopathological assessment was performed using H&E staining at 14 and 21 days postoperatively. The results showed that the double-layered scaffolds and nanofibers implanted with ADSCs significantly enhanced re-epithelialization, angiogenesis, and collagen remodeling of the wounds compared with the control group [57].

Electrospun fibers have potential applications in wound healing but need to possess antimicrobial, anti-inflammatory, and pro-angiogenic properties. Wang et al. improved their spinning properties by synthesizing sodium oxide selenobacterium alginate (OSA) and prepared composite fibrous membranes by mixing it with zinc oxide nanoparticles (ZnO-NPs). In vitro and in vivo studies showed that these membranes exhibited good biocompatibility, antimicrobial effect, and positive impact on wound healing. Such nanofiber membranes have a wide range of potential applications for wound healing [58]. Hu et al. developed a composite wound dressing consisting of alginate and PCL nanofibers. This dressing combines the moist environment of alginate, the cell adhesion enhancement of PCL, and the long-lasting antimicrobial properties of nanosilver. Meanwhile, plasmid DNA containing the platelet-derived growth factor-B (PDGF-B) gene was adsorbed onto the alginate fibers via carrier nanoparticles, enabling intracellular transfection on the wound and promoting wound healing. The release of calcium ions from the alginate fibers helped accelerate hemostasis. It was demonstrated that this composite dressing significantly improved the wound closure speed and promoted collagen formation, which had multifaceted healing-promoting advantages [59].

1.2.2 Tissue Regeneration

A composite scaffold composed of methacrylate alginate fiber and polycaprolactone fiber by electrospinning has been fabricated by Apsite et al. [60]. This scaffold can spontaneously curl into various forms in water and induce myoblasts to further differentiate into muscle tubes along the axial arrangement. It provides

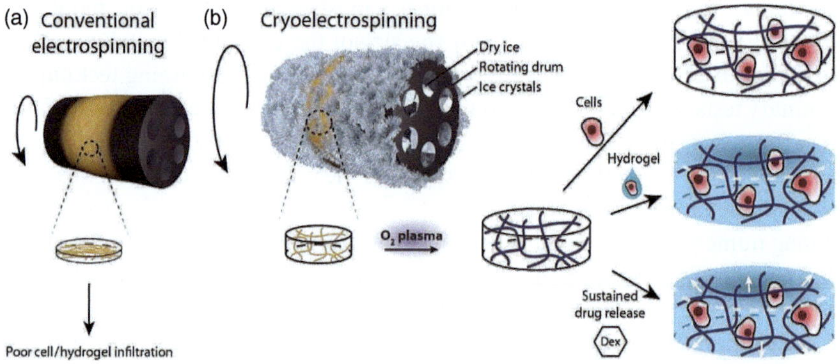

Figure 1.7 (a) Conventional rotary mandrel electrospinning forms a dense polymer network that prevents cell and hydrogel penetration. (b) Electrospinning on a −78 °C mandrel induces ice crystal co-deposition, which sublimates to form a super porous biocompatible membrane. Plasma treatment reduces the intrinsic hydrophobicity of PCL, making the skeleton permeable to cells or hydrogels. Source: Ref. [62]/John Wiley & Sons.

a feasible method for the preparation of artificial muscle. Yeo et al. [61] used 3D printing and electrospinning techniques to fabricate alginate-polycaprolactone scaffolds with uniaxial arrangement of micron and nanometer patterns. Due to the synergistic effect of different patterns, the expression of different muscle-derived genes in myoblasts was significantly increased. For simulating the components of natural extracellular matrix, Formica et al. [62] infiltrated chondrocytes/alginate solution into polycaprolactone fiber network, and then physically cross-linked to make cell-loaded electrospinning scaffolds. This scaffold can significantly induce chondrocytes to produce matrix rich in glycosaminoglycan and type II collagen and promote cartilage repair (Figure 1.7).

Hazeri et al. [63] further explored the effects of alginate sulfated PVA/alginate nanofibers with different concentrations (10, 20, and 30 wt%) on nerve regeneration. They found that nanofiber scaffolds containing 30 wt% alginate were suitable for the growth of bone marrow mesenchymal stem cells and induced bone marrow mesenchymal stem cells to differentiate into nerves over a period of up to 14 days.

Ductal carcinoma in situ (DCIS) is a breast cancer type, and Sadeghi et al. developed a three-layer tubular 3D scaffold based on a complex tissue environment for use in DCIS models and chemo-photothermal therapy studies. The scaffold consists of an intermediate layer of PLA/PCL nanofibers, and the outer and inner layers are dopamine alginate/PVA nanofibers, which provide a suitable environment for cell growth. The scaffolds were prepared by electrospinning to mimic the structure of natural extracellular matrix, and the inner and outer layers were added with dopamine nanoparticles with photothermal properties. MDA-MB-231 breast cancer cells were cultured on the scaffolds, and the synergistic effects of cell growth promotion and chemo-photothermal therapy were confirmed by in vitro experiments [64].

To enhance bone regeneration, Joo et al. prepared a sodium alginate-cast polycaprolactone-gelatin-β-calcium triphosphate bilayer membrane using an electrospinning process. This porous membrane degraded gradually in vivo, showing appropriate hydrophilicity and degradability. The electrospun membrane provided a favorable growth environment, while the alginate membrane inhibited cell attachment and was a nontoxic material. After implantation, the bilayer membrane promoted bone formation and effectively inhibited fibrous tissue infiltration. Immunocytochemical analysis showed that the bilayer membrane directed more proteins, controlled bone mineralization, and improved the guiding properties of tissue-engineered bone grafts [65].

Inflammation induced by myocardial infarction (MI) can lead to necrosis of cardiomyocytes. Hydrogen hydrosulfide (H_2S) is considered an important gas signaling molecule with various biological effects such as anti-inflammatory, antioxidant, and angiogenesis promotion. However, the feasibility of using H_2S directly to treat MI is limited by its short residence time and drastic side effects. To alleviate this problem, Li et al. investigated and developed a composite scaffold (AAB) that gradually releases H_2S from a patch made of modified alginate and albumin by electrospinning. The patch was exposed to an 808 nm laser, and the thermal energy generated by the black phosphorous nanosheets altered the molecular structure, enabling precise attachment to the myocardium. The AAB alleviated inflammation by reducing ROS levels and enhancing M2 macrophages. This engineered cardiac patch is expected to alleviate inflammation and promote angiogenesis after MI, thus helping to restore cardiac function and providing a new avenue for MI treatment [66] (Figure 1.8).

Wound healing is a complex biological process involving a series of cell signaling pathways and molecular interactions. In this process, microRNAs (miRNAs), as a class of short non-coding RNAs, play an important regulatory role. Bombin et al. found that down-regulation of microRNA-31 (miR-31) and microRNA-132 (miR-132) in wound healing was associated with delayed healing. Therefore, by effectively delivering these two microRNAs, it is expected to play an active role in promoting tissue repair. In order to fully utilize the efficacy of microRNAs, the delivery challenge must be addressed. They propose to employ RALA peptide nanocomplexes, which can encapsulate miR-31 and miR-132 into nanoparticles, to achieve efficient and safe intracellular delivery of these two microRNAs. By mixing the delivery of miR-31 and miR-132, the study aims to reduce potential toxicity while exploiting the synergistic effect of the two with a view to maximizing the therapeutic effect of promoting wound healing. As demonstrated by wound-healing experiments in mice, this technology significantly accelerated the rate of wound closure, increased epidermal thickness and blood vessel counts, demonstrating the potential for innovative therapeutic applications [67]. Chen et al. developed electrospun composite nanofibers of fucoidan derivatives using electrospinning technology to enhance the electrospun properties of fucoidan by chemically modifying the molecular flexibility of oxidized fucoidan derivatives (RAOA). These fibers were able to load the hydrophobic anti-inflammatory drug ibuprofen with the assistance of PVA. The synthesized RAOA was characterized by FT-IR, 1H NMR, and fluorescence photometry. The effects of surface tension, conductivity, and rheological properties of the RAOA/PVA blend solution on the electrospinning performance and fiber morphology were investigated. The drug loading and release mechanism of ibuprofen from RAOA/PVA electrospun

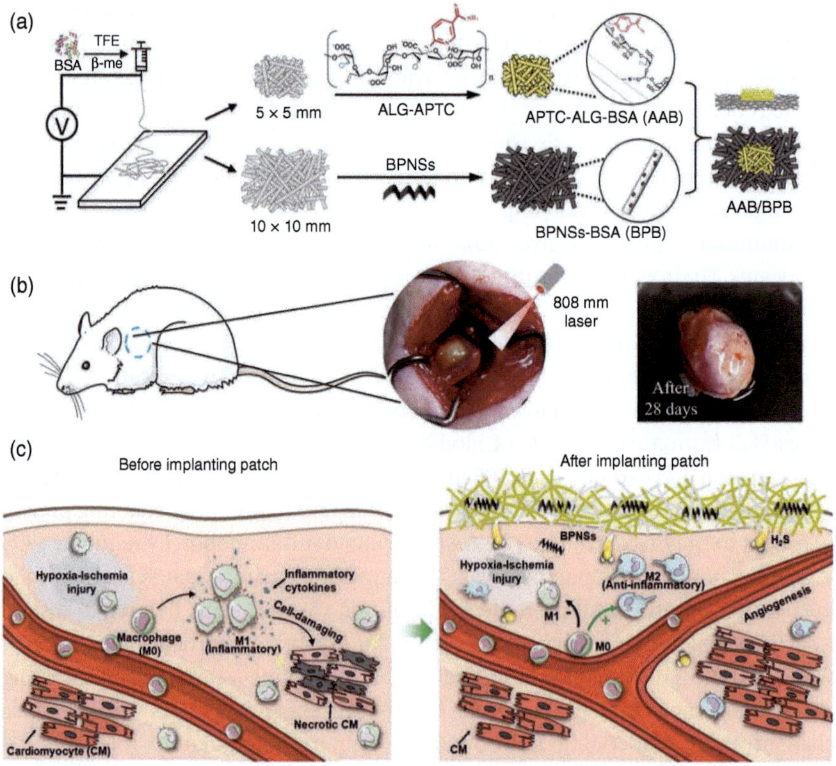

Figure 1.8 (a) Cardiac patch preparation process. (b) Application process: the patch is implanted into the infarcted area to cover and replace the dead tissue. (c) Mechanism of action: The implanted adMSC-secreted factor promotes neovascularization, and the hydrogel degrades to release Philadelphia chromatin-positive cells, which interact with host cells to improve cardiac function. Source: Ref. [66]/John Wiley & Sons/CC BY 4.0.

composite nanofibers was further analyzed by simulated drug release experiments. The RAOA/PVA electrospun composite nanofibers exhibited high encapsulation efficiency and sustainable-release properties, and low cytotoxicity to L929 cells. These electrospun nanofibers combine good self-assembly properties, colloidal interfacial activity, sustainable-release properties, and cytocompatibility, showing potential as functional wound dressings in biomedical applications [68]. Biodegradable and biocompatible scaffolds based on PVA/alginate mixtures were prepared by electrospinning followed by cross-linking with calcium chloride by Soto-Quintero et al. The morphology and degradation rate of the scaffolds were found to be tunable by varying the SA concentration (i.e. 3.5%, 4.0%, and 5.0% SA.) The highest SA content on the PVA/alginate scaffolds showed the highest degradation rate during the 100-day test period. In addition, these scaffolds experienced shrinkage and possible microstructural changes after degradation. The degradation of the scaffolds was studied by electrochemical impedance spectroscopy, and the lowest resistance and highest capacitance behaviors were found to correlate with high degradation rates, i.e. greater mass loss. Increasing the surface roughness of the scaffolds induced proper biocompatibility, as observed in 4.0% PVA/alginate scaffolds. Thus, HaCat cells in aqueous solution had a direct effect on the topology and stability (controlled degradation) of the scaffolds. These scaffolds have great potential in the field of tissue engineering, especially in cell regeneration [69].

Achieving functionally intact skin regeneration has long been a challenging task. As a multilayered and complex organ, the skin undergoes a continuous healing process influenced by multiple mechanisms. Critical nutrient, oxygen, and biochemical signals can direct specific cellular behaviors that ultimately contribute to the formation of high-quality tissue. Such biomolecular exchanges can be modulated through scaffold engineering, which is one of the frontiers in the field of skin substitution and equivalents. Molina et al. prepared a novel 3D fibrous scaffold consisting of PCL/sodium fucoxanthate (CA) aimed at inducing keratogenesis through the action of calcium loss. The scaffolds were prepared by electrospinning using a PCL/sodium fucoidan solution that was treated by immersion in a calcium chloride solution in order to replace the sodium ions attached to the fucoidan with calcium ions. This treatment not only provided ionic substitution but also induced cross-linking of the fibers. The in vitro performance of the scaffolds was investigated by growing and staining fibroblasts and keratin-forming cells on the scaffolds and using differentiation markers to detect the evolution of basal, spiny, and granular keratin-forming cells. The findings reveal the potential of PCL/CA scaffolds for tissue engineering and suggest that calcium loss in the scaffolds may contribute to the formation of a suitable biological environment that promotes the attachment, proliferation, and differentiation of major skin cells [70].

Ashraf et al. prepared collagen (Col)/sodium alginate (SA)/polyethylene oxide (PEO)/exocytidyl polysaccharide (EPS) nanofibers skin substitutes generated by *Rhodotorula mucilaginosa* sp. GUMS16 using biaxial electrostatic spinning technique. Among them, collagen is a natural scaffold, sodium alginate absorbs excess wound fluid, and GUMS16-generated EPS acts as an antioxidant. In this study, collagen and sodium alginate nanofibers containing different amounts of exocytopolysaccharide were successfully prepared by biaxial electrostatic spinning. The study aimed to improve the mechanical properties and cytocompatibility. The results showed uniform morphology, high hydrophilicity, cytocompatibility, and non-toxicity. Col-SA/PEO + EPS2% nanofibers showed superior mechanical properties and cellular behaviors and are expected to be potential cytocompatible scaffolds for various tissue engineering applications [71].

Building gene activation matrices (GAMs) by combining gene delivery with functional scaffolds is a promising strategy for tissue engineering. He et al. used nonviral DNA delivery nanocomplexes modified with PLGA/PEI nanoparticles carrying pVEGF plasmid DNA immobilized on electrospun fucoidan nanofibrous scaffolds to form GAMs. This innovative system exhibited low cytotoxicity, high transfection efficiency, and sustained gene release of VEGF was achieved in vitro. In a rat skin wound model, GAM accelerated wound healing, promoted re-epithelialization, attenuated inflammatory response, and enhanced neoangiogenesis. This study demonstrates the potential of the GAM system in tissue engineering [72]. Bioactive scaffolds for the treatment of large bone defects remain a great challenge in clinical applications. GAM, as a combination of gene therapy and tissue-engineered scaffolds, offers a promising strategy for the restoration of structure and function of damaged or dysfunctional tissues. He et al. developed a gene-activated bionic composite scaffold consisting of an outer sheath of electrospun polycaprolactone fibers and a hydrogel core of alginate carrying plasmid DNA encoding for bone-forming protein 2 (pBMP2) and vascular endothelial growth factor (pVEGF)) in an alginate hydrogel core. A low cytotoxic and efficient peptide-modified polymer nanocarrier was used as a nonviral DNA delivery vehicle. The obtained GAM enabled spatiotemporal release of pVEGF and pBMP2 and promoted osteogenic differentiation of preosteogenic osteoblasts in vitro. Through in vivo evaluation in a rat critical-size iliac defect model, the dual gene delivery system was demonstrated to significantly accelerate bone healing through activation of angiogenesis and osteogenesis. These results indicate that the developed fiber–hydrogel composite scaffolds with dual gene-activated nucleus-sheath structures have significant effects in the regeneration of critical-size bone defects, and confirm the potential of cell-free scaffold-based gene therapy in tissue engineering [73].

1.3 Alginate-Based 3D Printing for Biomedical Application

1.3.1 Alginate-Based Bio-Ink and Printing Strategies Improvement

3D printing technology can prepare various forms of biomaterials according to the actual needs, so it is widely used in the field of biomedicine. The improvement of 3D printing technology can be divided into two categories, one is to improve the properties of biological ink, the other is to improve printing strategies and methods.

Biological ink with natural polymer as the main component has poor stability in vivo, while synthetic polymer can cause immune rejection and have a certain toxic effect on cell proliferation. These problems make it a great challenge to reshape soft tissue through 3D printing technology. Van Belleghem et al. [74] innovatively combined natural polymers with synthetic polymers to prepare polyethylene glycol (PEG) covalently linked with naturally derived and physically cross-linked alginate double-network biological inks and biodegradable GelMA biological inks. The combined use of two kinds of biological ink can reduce the biological toxicity and ensure the structural stability at the same time. Jia et al. [75] improved the rheological properties and mechanical strength of biological ink by adding four-arm polyethylene glycol-tetraacrylate (PEGTA) to gelatin methacryloyl (GelMA) and sodium alginate biological ink. Firstly, Ca^{2+} was pre-cross-linked with alginate to construct the basic framework, and then the whole structure was permanently shaped by the photocross-linking of GelMA and 4-arm PEGTA components. Four-arm PEGTA has a number of active cross-linking sites to significantly improve the stability of the structure, coupled with the use of three-layer coaxial nozzles, they successfully prepared a perfusion vascular structure. Pataky et al. [76] invented a new 3D printing strategy using the rapid cross-linking of alginate with Ca^{2+}. They used the hydrated gelatin matrix as the Ca^{2+} repository and used the drop-by-drop printing method to make the alginate biological ink accumulate like a wall to finely control the shape of the printing structure. Kirillova et al. [77] used 4D bio-printing technology to prepare vascular structure. They first modified alginate with methacrylate group to make it have the ability of photocross-linking, then printed alginate film on glass, and then curled the film spontaneously to form vascular poplar structure. The vascular poplar structure with a diameter of 20–150 μm can be prepared by controlling the thickness of the film.

The rapid prototyping performance of 3D printing bioinks is crucial for printing high-resolution fine structures, and Jeon et al. successfully achieved the preparation of complex pendant structures without external support by modifying alginate materials with methacrylated materials that utilize shear-thinning and calcium cross-linking properties. Thanks to its excellent biocompatibility, this bioink can be used for stem cell loading and organoid preparation [78].

In tissue engineering, the preparation of cellular scaffolds containing hollow microchannels that resemble blood vessels is critical for building thick cellular tissues. However, existing techniques have limited success in producing hollow channels with diameters smaller than a few hundred micrometers, and Bolívar-Monsalve et al. prepared hollow microchannels with cellular widths in a single step by co-extruding a methacryloyl-alginate hydrogel suspended from C2C12 of murine myoblasts and a sacrificial material, hydroxyethylcellulose, using a printhead with Kenics static mixing elements, and vascularized skeletal muscle-like fibrous scaffolds. Compared with solid scaffolds, the viability and metabolic activity of C2C12 myoblasts were higher in the hollow scaffolds. In addition, the hollow channels alleviated hypoxia, promoted the expression of Ki67, induced 82% of the cells to orient themselves, and facilitated the rapid elongation, differentiation, and maturation of the C2C12 cells into myogenic fibers [79].

The gradual weakening of bioelectrical stimulation around the wound leads to the downregulation of the wound-healing cascade and disorders in collagen fiber deposition, which is the main cause of delayed wound healing and scar tissue formation. The ZnO nanoparticle-modified PVDF/SA piezoelectric hydrogel scaffolds (ZPFSA), which were prepared by three-dimensional printing by Liang et al., were designed to solve the problem of gradual weakening of bioelectrical stimulation in the course of prolonged wound healing, diminishing during prolonged wound repair. The scaffold has a dual piezoresponse model that mimics and amplifies endogenous bioelectricity to promote wound healing and prevent scar formation. ZPFSA significantly accelerated wound healing within two weeks and prevented scar formation by modulating cell migration and angiogenic cascade responses [80].

1.3.2 Attempts at Bionic Matrix Ink

Extracellular matrix (ECM) has attracted much attention in promoting tissue regeneration, but its slow gel kinetics limits its application in 3D printing. In order to solve this problem, De Santis et al. [81] turned their attention to alginate which can be rapidly cross-linked with divalent cations. The tubes and branching structures

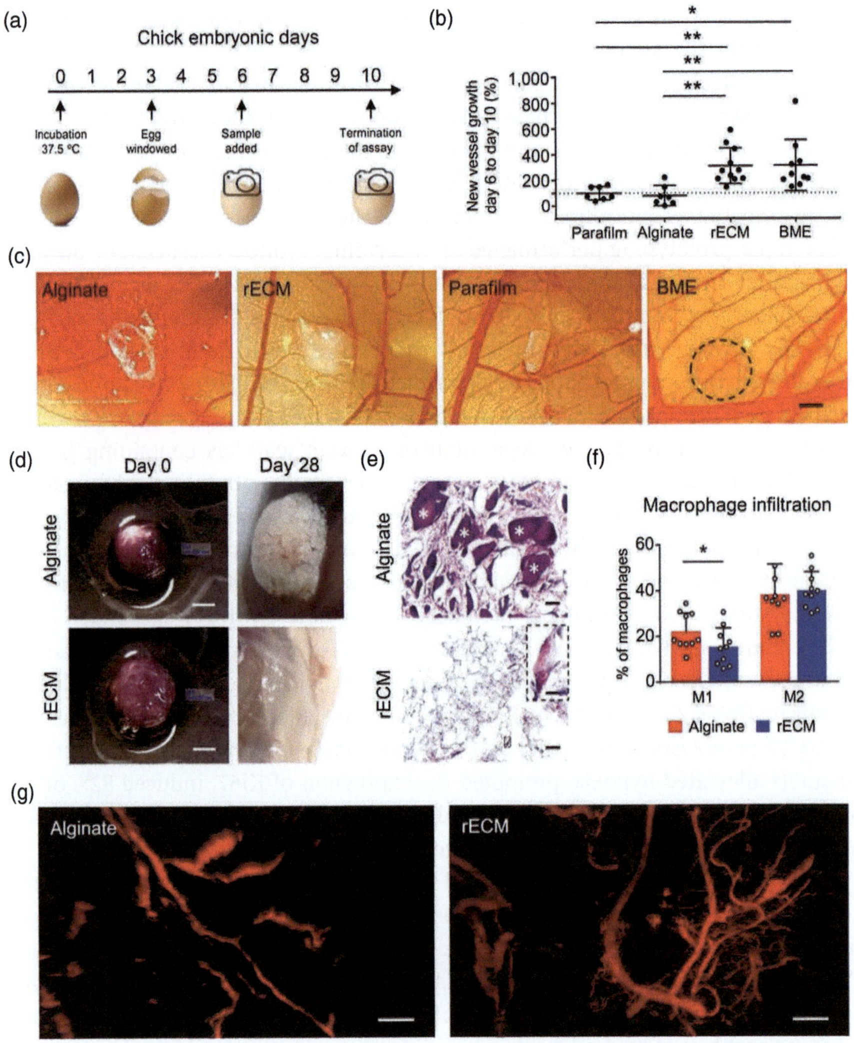

Figure 1.9 (a) Overview of CAM experiments. (b) Quantitative comparison of vascular growth induced by different hydrogels. rECM hydrogel induces better vascular growth than sodium alginate hydrogel. (c) Schematic diagram of CAM samples. rECM hydrogel is surrounded by abundant blood vessels. (d) Shape changes of different hydrogels implanted subcutaneously. (e) HE staining after 28 days of implantation. rECM hydrogel showed a few non-protein fragments, and both hydrogels showed neovascularization. (f) Inflammatory cell infiltration was lower in the early stage of rECM hydrogel implantation. (g) Spontaneous fluorescence showed better vascular regeneration in rECM hydrogel. Source: Ref. [81]/ John Wiley & Sons/CC BY 4.0.

prepared by the mixed biological ink system enhanced by alginate and extracellular matrix have been proved to promote cell proliferation and differentiation and angiogenesis in vivo (Figure 1.9).

Xie et al. [82] explored the potential application of 3D printing technology in drug screening in vitro. They successfully constructed a liver cancer model in a 3D structure made from a mixture of gelatin and sodium alginate. In the process of long-term culture in vitro, its tumorigenic potential and histological characteristics were preserved, which provided a platform for screening anticancer drugs in vitro.

Ma et al. constructed a multi-scale hierarchical bioactive calcium silicate nanowire/alginate composite hydrogel scaffold for tendon-to-bone interface tissue engineering. Three-dimensional printing technology was combined with mechanical stretching methods to introduce biomimetic reinforcement structures from nano- to micrometer- to micro-scale in this composite hydrogel, which significantly improved the mechanical properties. Experiments showed that the biochemical and topographical characteristics of the composite hydrogel provided a favorable microenvironment for rabbit bone marrow mesenchymal stem cells and tendon stem cells to promote their directional alignment and induced differentiation. The composite scaffold significantly promoted the regeneration of tendon–bone tissue in vivo, especially in the fibrocartilage transition zone. Therefore, this multi-scale structural design provides an innovative strategy for the engineering of functionalized tendon–bone tissues [83].

Zhu et al. prepared a multifunctional nanocomposite bioink for extrusion bioprinting using amine-functionalized copper (Cu), oxidized alginate, gelatin, and mesoporous bioactive glass nanoparticles (ACuMBGNs). This ink has good rheological properties and structural stability. Rapid cell spreading and high survival were supported by the reversible dynamic microenvironment and cell adhesion ligands introduced by aminated particles. Osteogenic differentiation and angiogenesis of mouse bone marrow stromal stem cells (BMSCs) were promoted in the bioprinted scaffolds without additional growth factors. This nanocomposite biomaterial is expected to be a superior platform for bioprinting complex three-dimensional matrix environments for superior cell support [84].

References

1 Dekamin, M.G., Karimi, Z., Latifidoost, Z. et al. (2018). Alginic acid: a mild and renewable bifunctional heterogeneous biopolymeric organocatalyst for efficient and facile synthesis of polyhydroquinolines. *Int. J. Biol. Macromol.* 108: 1273–1280.

2 Lee, K.Y. and Mooney, D.J. (2012). Alginate: properties and biomedical applications. *Prog. Polym. Sci.* 37: 106–126.

3 Cao, L.Q., Lu, W., Mata, A. et al. (2020). Egg-box model-based gelation of alginate and pectin: a review. *Carbohydr. Polym.* 242: 116389.

4 Czuban, M., Srinivasan, S., Yee, N.A. et al. (2018). Bio-orthogonal chemistry and reloadable biomaterial enable local activation of antibiotic prodrugs and enhance treatments against infections. *ACS Cent. Sci.* 4: 1624–1632.

5 Ke, Y.H., Zhu, J.M., Chu, Y.H. et al. (2022). Bifunctional fusion membrane-based hydrogel enhances antitumor potency of autologous cancer vaccines by activating dendritic cells. *Adv. Funct. Mater.* 32: 2201306.

6 Zhao, Z.P., Li, Q., Qin, X.H. et al. (2022). An injectable hydrogel reshaping adenosinergic axis for cancer therapy. *Adv. Funct. Mater.* 32: 2200801.

7 Wu, Y.H., Yao, Y., Zhang, J.M. et al. (2022). Tumor-targeted injectable double-network hydrogel for prevention of breast cancer recurrence and wound infection via synergistic photothermal and brachytherapy. *Adv. Sci.* 9: 2200681.

8 Li, Z.T., Ding, Y.Y., Liu, J. et al. (2022). Depletion of tumor associated macrophages enhances local and systemic platelet-mediated anti-PD-1 delivery for post-surgery tumor recurrence treatment. *Nat. Commun.* 13: 1845.

9 Zhu, Z.Q., Wang, L., Peng, Y.O. et al. (2022). Continuous self-oxygenated double-layered hydrogel under natural light for real-time infection monitoring, enhanced photodynamic therapy, and hypoxia relief in refractory diabetic wounds healing. *Adv. Funct. Mater.* 32: 2201875.

10 Gan, J.J., Sun, L.Y., Chen, G.P. et al. (2022). Mesenchymal stem cell exosomes encapsulated oral microcapsules for acute colitis treatment. *Adv. Healthc. Mater.* 11: e2201105.

11 Zhang, D., Li, Q., Chen, X.W. et al. (2022). An injectable hydrogel to modulate T cells for cancer immunotherapy. *Small* 18: e2202663.

12 Delcassian, D., Luzhansky, I., Spanoudaki, V. et al. (2020). Magnetic retrieval of encapsulated beta cell transplants from diabetic mice using dual-function MRI visible and retrievable microcapsules. *Adv. Mater.* 32: 1904502.

13 Hasani-Sadrabadi, M.M., Sarrion, P., Pouraghaei, S. et al. (2020). An engineered cell-laden adhesive hydrogel promotes craniofacial bone tissue regeneration in rats. *Sci. Transl. Med.* 12: eaay6853.

14 Whitehead, J., Griffin, K.H., Gionet-Gonzales, M. et al. (2021). Hydrogel mechanics are a key driver of bone formation by mesenchymal stromal cell spheroids. *Biomaterials* 269: 120607.

15 Hung, B., Gonzalez-Fernandez, T., Lin, J.B. et al. (2020). Multi-peptide presentation and hydrogel mechanics jointly enhance therapeutic duo-potential of entrapped stromal cells. *Biomaterials* 245: 119973.

16 Hasturk, O., Smiley, J.A., Arnett, M. et al. (2022). Cytoprotection of human progenitor and stem cells through encapsulation in alginate templated, dual crosslinked silk and silk-gelatin composite hydrogel microbeads. *Adv. Healthc. Mater.* 11: e2200293.

17 Qian, B., Yang, Q., Wang, M.L. et al. (2022). Encapsulation of lyophilized platelet-rich fibrin in alginate-hyaluronic acid hydrogel as a novel vascularized substitution for myocardial infarction. *Bioact. Mater.* 7: 401–411.

18 Panebianco, C.J., Rao, S.J., Hom, W.W. et al. (2022). Genipin-crosslinked fibrin seeded with oxidized alginate microbeads as a novel composite biomaterial strategy for intervertebral disc cell therapy. *Biomaterials* 287: 121641.

19 Theocharidis, G., Rahmani, S., Lee, S.M. et al. (2022). Murine macrophages or their secretome delivered in alginate dressings enhance impaired wound healing in diabetic mice. *Biomaterials* 288: 121692.

20 Jeon, O. and Alsberg, E. (2013). Regulation of stem cell fate in a three-dimensional micropatterned dual-crosslinked hydrogel system. *Adv. Funct. Mater.* 23: 4765–4775.

21 Gonzalez-Pujana, A., Vining, K.H., Zhang, D.K.Y. et al. (2020). Multifunctional biomimetic hydrogel systems to boost the immunomodulatory potential of mesenchymal stromal cells. *Biomaterials* 257: 120266.

22 Liu, H.T., Wang, Y.Q., Wang, H. et al. (2020). A droplet microfluidic system to fabricate hybrid capsules enabling stem cell organoid engineering. *Adv. Sci.* 7: 1903739.

23 Liao, J.F., Tian, T.R., Shi, S.R. et al. (2017). The fabrication of biomimetic biphasic CAN-PAC hydrogel with a seamless interfacial layer applied in osteochondral defect repair. *Bone Res.* 5: 17018.

24 Öztürk, E., Arlov, O., Aksel, S. et al. (2016). Sulfated hydrogel matrices direct mitogenicity and maintenance of chondrocyte phenotype through activation of FGF signaling. *Adv. Funct. Mater.* 26: 3649–3662.

25 Liao, I.C., Moutos, F.T., Estes, B.T. et al. (2013). Composite three-dimensional woven scaffolds with interpenetrating network hydrogels to create functional synthetic articular cartilage. *Adv. Funct. Mater.* 23: 5833–5839.

26 Moshaverinia, A., Xu, X.T., Chen, C. et al. (2014). Application of stem cells derived from the periodontal ligament or gingival tissue sources for tendon tissue regeneration. *Biomaterials* 35: 2642–2650.

27 Mredha, M.T.I., Guo, Y.Z., Nonoyama, T. et al. (2018). A facile method to fabricate anisotropic hydrogels with perfectly aligned hierarchical fibrous structures. *Adv. Mater.* 30: 1870060.

28 Campbell, K.T., Stilhano, R.S., and Silva, E.A. (2018). Enzymatically degradable alginate hydrogel systems to deliver endothelial progenitor cells for potential revasculature applications. *Biomaterials* 179: 109–121.

29 Hao, T., Li, J.J., Yao, F.L. et al. (2018). Injectable fullerenol/alginate hydrogel for suppression of oxidative stress damage in Brown adipose-derived stem cells and cardiac repair (vol 11, pg 5474, 2017). *ACS Nano* 12: 10564.

30 Anker, S.D., Coats, A.J.S., Cristian, G. et al. (2015). A prospective comparison of alginate-hydrogel with standard medical therapy to determine impact on functional capacity and clinical outcomes in patients with advanced heart failure (AUGMENT-HF trial). *Eur. Heart J.* 36: 2297–2309.

31 Leor, J., Tuvia, S., Guetta, V. et al. (2009). Intracoronary injection of in situ forming alginate hydrogel reverses left ventricular remodeling after myocardial infarction in swine. *J. Am. Coll. Cardiol.* 54: 1014–1023.

32 He, Y.N., Li, F., Jiang, P. et al. (2023). Remote control of the recruitment and capture of endogenous stem cells by ultrasound for in situ repair of bone defects. *Bioact. Mater.* 21: 223–238.

33 Zheng, A., Wang, X., Xin, X.Z. et al. (2023). Promoting lacunar bone regeneration with an injectable hydrogel adaptive to the microenvironment. *Bioact. Mater.* 21: 403–421.

34 Wang, C.R., Jiang, X., Kim, H.J. et al. (2022). Flexible patch with printable and antibacterial conductive hydrogel electrodes for accelerated wound healing. *Biomaterials* 285: 121479.

35 Wang, T., Yi, W.W., Zhang, Y. et al. (2023). Sodium alginate hydrogel containing platelet-rich plasma for wound healing. *Colloids Surf. B* 222: 113096.

36 Choi, S., Moon, J.R., Park, N. et al. (2023). Bone-adhesive anisotropic tough hydrogel mimicking tendon enthesis. *Adv. Mater.* 35: 2206207.

37 Liu, H.S., Hu, X.L., Li, W. et al. (2023). A highly-stretchable and adhesive hydrogel for noninvasive joint wound closure driven by hydrogen bonds. *Chem. Eng. J.* 452: 139368.

38 Yao, S.S., Zhao, Y.Q., Xu, Y.F. et al. (2022). Injectable dual-dynamic-bond cross-linked hydrogel for highly efficient infected diabetic wound healing. *Adv. Healthc. Mater.* 11: e2200516.

39 Wang, C.P., Wang, X.Y., Dong, K.Y. et al. (2016). Injectable and responsively degradable hydrogel for personalized photothermal therapy. *Biomaterials* 104: 129–137.

40 Patrick, P.S., Bear, J.C., Fitzke, H.E. et al. (2020). Radio-metal cross-linking of alginate hydrogels for non-invasive imaging. *Biomaterials* 243: 119930.

41 Zhu, Y.L., Ma, Z.J., Kong, L.Z. et al. (2020). Modulation of macrophages by bioactive glass/sodium alginate hydrogel is crucial in skin regeneration enhancement. *Biomaterials* 256: 120216.

42 Ren, J.L., Yin, X.J., Chen, Y. et al. (2020). Alginate hydrogel-coated syringe needles for rapid haemostasis of vessel and viscera puncture. *Biomaterials* 249: 120019.

43 Jons, C.K., Grosskopf, A.K., Baillet, J. et al. (2022). Yield-stress and creep control depot formation and persistence of injectable hydrogels following subcutaneous administration. *Adv. Funct. Mater.* 32: 2203402.

44 Ji, D., Park, J.M., Oh, M.S. et al. (2022). Superstrong, superstiff, and conductive alginate hydrogels. *Nat. Commun.* 13: 3019.

45 Hwang, J.C., Kim, M., Kim, S. et al. (2022). In situ diagnosis and simultaneous treatment of cardiac diseases using a single-device platform. *Sci. Adv.* 8: abq0897.

46 Asadi-Korayem, M., Akbari-Taemeh, M., Mohammadian-Sabet, F. et al. (2021). How does counter-cation substitution influence inter- and intramolecular hydrogen bonding and electrospinnability of alginates. *Int. J. Biol. Macromol.* 171: 234–241.

47 Fujita, S., Wakuda, Y., Matsumura, M., and Suye, S.I. (2019). Geometrically customizable alginate hydrogel nanofibers for cell culture platforms. *J. Mater. Chem. B* 7: 6556–6563.

48 Kaassis, A.Y.A., Young, N., Sano, N. et al. (2014). Pulsatile drug release from electrospun poly(ethylene oxide)-sodium alginate blend nanofibres. *J. Mater. Chem. B* 2: 1400–1407.

49 De Silva, R.T., Dissanayake, R.K., Mantilaka, M.M.M.G.P.G. et al. (2018). Drug-loaded halloysite nanotube-reinforced electrospun alginate-based nanofibrous scaffolds with sustained antimicrobial protection. *ACS Appl. Mater. Interfaces* 10: 33913–33922.

50 Dodero, A., Alloisio, M., Castellano, M., and Vicini, S. (2020). Multilayer alginate-polycaprolactone electrospun membranes as skin wound patches with drug delivery abilities. *ACS Appl. Mater. Interfaces* 12: 31162–31171.

51 Mulholland, E.J., Ali, A., Robson, T. et al. (2019). Delivery of RALA/siFKBPL nanoparticles electrospun bilayer nanofibres: an innovative angiogenic therapy for wound repair. *J. Control. Release* 316: 53–65.

52 Tang, Y.D., Lan, X.Z., Liang, C.F. et al. (2019). Honey loaded alginate/PVA nanofibrous membrane as potential bioactive wound dressing. *Carbohydr. Polym.* 219: 113–120.

53 Hajiali, H., Summa, M., Russo, D. et al. (2016). Alginate-lavender nanofibers with antibacterial and anti-inflammatory activity to effectively promote burn healing. *J. Mater. Chem. B* 4: 1686–1695.

54 Dodero, A., Scarfi, S., Pozzolini, M. et al. (2020). Alginate-based electrospun membranes containing ZnO nanoparticles as potential wound healing patches: biological, mechanical, and physicochemical characterization. *ACS Appl. Mater. Interfaces* 12: 3371–3381.

55 Chen, Y.H., Cheng, C.H., Chang, W.J. et al. (2016). Studies of magnetic alginate-based electrospun matrices crosslinked with different methods for potential hyperthermia treatment. *Mater. Sci. Eng. C* 62: 338–349.

56 Alloisio, M., Dodero, A., Alberti, S. et al. (2022). Electrospun alginate mats embedding silver nanoparticles with bioactive properties. *Int. J. Biol. Macromol.* 213: 427–434.

57 Lashkari, M., Rahmani, M., Yousefpoor, Y. et al. (2023). Cell-based wound dressing: bilayered PCL/gelatin nanofibers-alginate/collagen hydrogel scaffold loaded with mesenchymal stem cells. *Int. J. Biol. Macromol.* 239: 124099.

58 Wang, W., Liu, M.Y., Shafiq, M. et al. (2023). Synthesis of oxidized sodium alginate and its electrospun bio-hybrids with zinc oxide nanoparticles to promote wound healing. *Int. J. Biol. Macromol.* 232: 123480.

59 Hu, W.W. and Lin, Y.T. (2022). Alginate/polycaprolactone composite fibers as multifunctional wound dressings. *Carbohydr. Polym.* 289: 119440.

60 Apsite, I., Uribe, J.M., Posada, A.F. et al. (2020). 4D biofabrication of skeletal muscle microtissues. *Biofabrication* 12: 015016.

61 Yeo, M. and Kim, G. (2019). Nano/microscale topographically designed alginate/PCL scaffolds for inducing myoblast alignment and myogenic differentiation. *Carbohydr. Polym.* 223: 115041.

62 Formica, F.A., Öztürk, E., Hess, S.C. et al. (2016). A bioinspired ultraporous nanofiber-hydrogel mimic of the cartilage extracellular matrix. *Adv. Healthc. Mater.* 5: 3129–3138.

63 Hazeri, Y., Irani, S., Zandi, M., and Pezeshki-Modaress, M. (2020). Polyvinyl alcohol/sulfated alginate nanofibers induced the neuronal differentiation of human bone marrow stem cells. *Int. J. Biol. Macromol.* 147: 946–953.

64 Sadeghi, M., Falahi, F., Akbari-Birgani, S., and Nikfarjam, N. (2023). Trilayer tubular scaffold to mimic ductal carcinoma breast cancer for the study of chemo-photothermal therapy. *ACS Appl. Polym. Mater.* 5: 2394–2407.

65 Joo, G., Park, M., Park, S.S. et al. (2022). Tailored alginate/PCL-gelatin-β-TCP membrane for guided bone regeneration. *Biomed. Mater.* 17: 045011.

66 Li, W.R., Chen, P.E., Pan, Y.X. et al. (2022). Construction of a band-aid like cardiac patch for myocardial infarction with controllable HS release. *Adv. Sci.* 9: e2204509.

67 Bombin, A.D.J., Dunne, N., and McCarthy, H.O. (2023). Delivery of a peptide/microRNA blend electrospun antimicrobial nanofibres for wound repair. *Acta Biomater.* 155: 304–322.

68 Chen, X.Q., Zhu, Q.M., Wen, Y.S. et al. (2022). Chemical modification of alginate via the oxidation-reductive amination reaction for the development of alginate derivative electrospun composite nanofibers. *J. Drug Deliv. Sci. Technol.* 68: 103113.

69 Soto-Quintero, A., González-Alva, P., Covelo, A., and Hernández, M.A. (2022). Study of the in vitro degradation and characterization of the HaCat keratinocytes adherence on electrospun scaffolds based polyvinyl alcohol/sodium alginate. *J. Appl. Polym. Sci.* 139: e52775.

70 Molina, M.I.E., Chen, C.A., Martinez, J. et al. (2023). Novel electrospun polycaprolactone/calcium alginate scaffolds for skin tissue engineering. *Materials* 16: 136.

71 Ashraf, S.S., Parivar, K., Roodbari, N.H. et al. (2022). Fabrication and characterization of biaxially electrospun collagen/alginate nanofibers, improved with *Rhodotorula mucilaginosa* sp. GUMS16 produced exopolysaccharides for wound healing applications. *Int. J. Biol. Macromol.* 196: 194–203.

72 He, S., Fang, J., Zhong, C.X. et al. (2022). Controlled pVEGF delivery via a gene-activated matrix comprised of a peptide-modified non-viral vector and a nanofibrous scaffold for skin wound healing. *Acta Biomater.* 140: 149–162.

73 He, S., Fang, J., Zhong, C.X. et al. (2022). Spatiotemporal delivery of pBMP2 and pVEGF by a core-sheath structured fiber-hydrogel gene-activated matrix loaded with peptide-modified nanoparticles for critical-sized bone defect repair. *Adv. Healthc. Mater.* 11: e2201096.

74 Van Belleghem, S., Torres, L., Santoro, M. et al. (2020). Hybrid 3D printing of synthetic and cell-laden bioinks for shape retaining soft tissue grafts. *Adv. Funct. Mater.* 30: 1907145.

75 Jia, W.T., Gungor-Ozkerim, P.S., Zhang, Y.S. et al. (2016). Direct 3D bioprinting of perfusable vascular constructs using a blend bioink. *Biomaterials* 106: 58–68.

76 Pataky, K., Braschler, T., Negro, A. et al. (2012). Microdrop printing of hydrogel bioinks into 3D tissue-like geometries. *Adv. Mater.* 24: 391–396.

77 Kirillova, A., Maxson, R., Stoychev, G. et al. (2017). 4D biofabrication using shape-morphing hydrogels. *Adv. Mater.* 29: 1703443.

78 Jeon, O., Bin Lee, Y., Lee, S.J. et al. (2022). Stem cell-laden hydrogel bioink for generation of high resolution and fidelity engineered tissues with complex geometries. *Bioact. Mater.* 15: 185–193.

79 Bolívar-Monsalve, E.J., Ceballos-González, C.F., Chávez-Madero, C. et al. (2022). One-step bioprinting of multi-channel hydrogel filaments using chaotic advection: fabrication of pre-vascularized muscle-like tissues. *Adv. Healthc. Mater.* 11: e2200448.

80 Liang, J.C., Zeng, H.J., Qiao, L. et al. (2022). 3D printed piezoelectric wound dressing with dual piezoelectric response models for scar-prevention wound healing. *ACS Appl. Mater. Interfaces* 14: 30507–30522.

81 De Santis, M.M., Alsafadi, H.N., Tas, S. et al. (2021). Extracellular-matrix-reinforced bioinks for 3D bioprinting human tissue. *Adv. Mater.* 33: 2005476.

82 Xie, F.H., Sun, L.J., Pang, Y. et al. (2021). Three-dimensional bio-printing of primary human hepatocellular carcinoma for personalized medicine. *Biomaterials* 265: 120416.

83 Ma, H.S., Yang, C., Ma, Z.J. et al. (2022). Multiscale hierarchical architecture-based bioactive scaffolds for versatile tissue engineering. *Adv. Healthc. Mater.* 11: 2102837.

84 Zhu, H., Monavari, M., Zheng, K. et al. (2022). 3D bioprinting of multifunctional dynamic nanocomposite bioinks incorporating Cu-doped mesoporous bioactive glass nanoparticles for bone tissue engineering. *Small* 18: 2104996.

2

Sources, Structures, and Properties of Cellulose

Cellulose is the most abundant natural linear polysaccharide, which is widely found in wood, algae, bacteria, and fungi. French scientist Auselms Payen first discovered and put forward the concept of cellulose in 1837, and bacterial cellulose (BC) was first discovered by A.J. Brown in 1886. BC is mainly produced by Gram-negative bacteria (*Glucoacetobacterxylinum* is the mainstream cellulose-producing bacteria). Compared with plant cellulose, BC has higher mechanical properties and porosity.

Cellulose $(C_6H_{10}O_5)_n$ is a polymer composed of D-glucose units, which is connected to each other by β-(1-4) glycosidic bonds [1]. Each glucose unit contains three hydroxyl groups, and strong intermolecular hydrogen bonds make cellulose a rigid long-chain fiber insoluble in water. At present, cellulose can be divided into four main types: cellulose I, cellulose II, cellulose III, cellulose IV, in which cellulose I is the main natural existing form. Cellulose I can be converted into cellulose II by alkali treatment or solubilization and crystallization, and cellulose III can be obtained by liquid nitrogen treatment, and cellulose IV can be prepared by glycerol heating. Due to the low immunogenicity, high mechanical strength, and biodegradability of cellulose, coupled with its wide range of biological sources and low price, its application potential in the biomedical field has been widely tapped. However, due to the low solubility of cellulose, its hydrolysate, D-glucose, is difficult to absorb in human body, which restricts its further application [2].

At present, the modification of cellulose materials mainly revolves around the three hydroxyl groups of cellulose, and the properties of cellulose-based materials are improved by substitution reaction or oxidation reaction to prepare cellulose ether or cellulose ester.

2.1 Cellulose-Based Hydrogel for Biomedical Application

2.1.1 Drug Delivery

The existence of the blood–brain barrier prevents most drugs from reaching the brain. For this reason, Wang et al. [3] took another approach to increase the

Natural Polymers for Biomedical Applications, First Edition. Wenguo Cui and Lei Xiang.

blood drug concentration in the brain by administering the scalp cortex. They prepared hyaluronic acid and methylcellulose-injectable hydrogels loaded with erythropoietin, using fine needles to achieve local drug release in the brain ventricles. The stroke cavity was significantly reduced by drug delivery 4 and 11 days after stroke, and the number of neurons in the peri-infarction area and migrating neuroblasts in the subventricular area was significantly increased. Appel et al. [4] prepared viologen-cucurbit[8]uril-naphthyl moiety ternary complex according to the property of cucurbit[8]uril that can accommodate two guests and added cellulose to make a hydraulic gel with extremely high water content. Glue. The hydrogel containing only 1.5 wt% polymer component can achieve sustained release of bovine serum albumin within 160 days.

Zhang et al. investigated and designed a novel cellulose-based biodegradable hydrogel material with self-healing function. They introduced a boronic acid group onto CMC backbone by chemical coupling and cross-linked it with PVA through reversible boronate dynamic bonding to form a hydrogel. The hydrogel was easy and efficient to synthesize, rapid in gelation, and could be locally injected for irregular wounds with good adhesion to tissues. The hydrogel can effectively reduce wound bleeding and significantly accelerate the healing of burn wounds. As a controlled release system, it can reduce the toxicity of the drug and maintain a good anti-tumor effect [5].

BC, a polysaccharide produced by microorganisms with high water absorption and biocompatibility, is an excellent raw material for wound dressings, but the lack of bactericidal activity cannot meet the requirements of wound healing. To remedy this deficiency, Zhao et al. introduced ursolic acid with antimicrobial effect and constructed a novel hydrogel membrane by Sanxan gel. This composite membrane exhibited excellent sustained drug release, antimicrobial eradication, biofilm inhibition, and superior blood and cytocompatibility. In a mouse wound infection model, the composite membrane promoted wound healing and attenuated inflammation and bacterial infection [6]. For the treatment of diabetes-induced difficult-to-heal chronic skin wounds, Zhang et al. designed a novel hydrogel dressing. The gel was prepared by cross-linking L-arginine-modified carboxymethylcellulose and chitosan, which had the dual functions of antibacterial and NO release. The gel mesh structure with small pore size could inhibit bacterial invasion and provide a moist microenvironment to promote healing, and the continuous release of NO could stimulate angiogenesis and reduce ischemia. Animal experiments showed that the gel dressing could significantly enhance the granulation tissue growth, collagen deposition, and vascular regeneration of diabetic ulcers without obvious toxic side effects [7].

Hydrogels containing quaternary ammonium salt (QAS) moieties perform well in acute wound treatment, but cytotoxicity and adhesion issues limit their application. Zhong et al. developed an elutable adaptive dressing using a dynamic layer of cellulose sulfate (CS) coated on a QAS hydrogel. In the early-wound stage, the CS coating released QAS for maximum disinfection; when the wound gradually healed, the CS layer remained stable, shielded QAS, and promoted epithelial regeneration. At the same time, the CS modulates the properties of the hydrogel, giving the dressing excellent wound sealing and hemostatic properties. This smart dressing based on

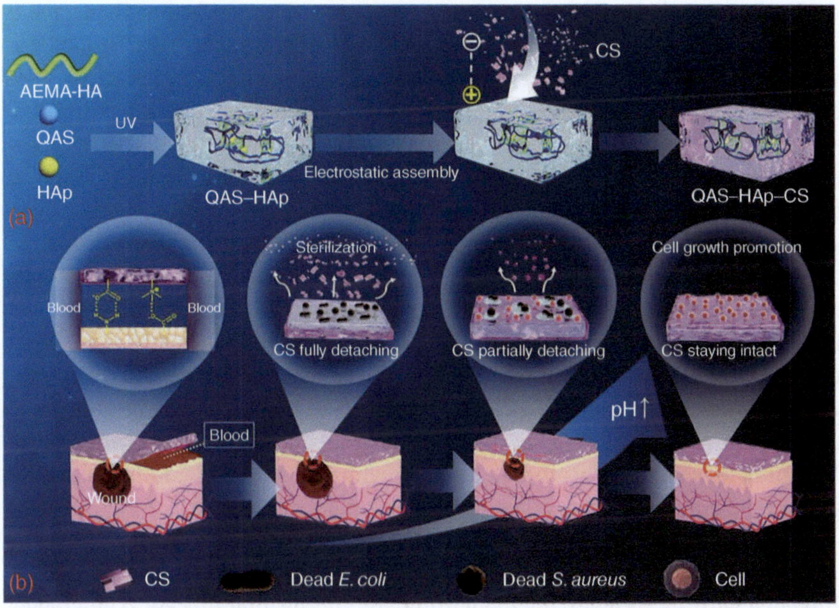

Figure 2.1 (a) Preparation of QAS–HAp–CS hydrogel: QAS monomer was first polymerized with cross-linking agent AEMA-HA in HAp solution to form QAS–HAp hydrogel under ultraviolet light irradiation, and then it was immersed in CS solution for coating to obtain QAS–HAp–CS hydrogel. (b) Mechanism of action of QAS–HAp–CS hydrogel as wound dressing: The coating makes the hydrogel surface hydrophobic, which can drain blood from the wound to reduce the interfacial gap and improve the adhesion through hydrogen bonding interactions. In the early stage of wound healing, the pH was about 5.5, and the acidic environment caused the CS coating to peel off, exposing the positively charged QAS groups for effective sterilization. As the wound healed, the pH rose to about 7.5, the CS coating remained intact, and most of the QAS groups were shielded, which facilitated cell proliferation. Source: Ref. [8]/John Wiley & Sons.

dynamic molecular interactions has the potential for a wide range of applications in medical therapy and health monitoring [8] (Figure 2.1).

Liu et al. designed a multifunctional injectable supramolecular gel wound dressing. The gel material achieved precise regulation of gel properties by constructing a dynamic and reversible network structure. They used the assembly of Pluronic F127 with different terminal groups and modified chitosan to form the main network and introduced nanocellulose to enhance the mechanical strength, and formed physical cross-linking points through self-assembly. The gel material exhibited excellent biocompatibility, antimicrobial, and self-repairing properties. The drug-loaded release could be regulated by light, pH, and temperature stimuli. Animal studies confirmed that the gel could significantly promote wound healing [9].

Cellulose hydrogels are highly regarded for their biological properties in wound healing. Zhang et al. prepared a composite hydrogel of bamboo marrow cellulose and carboxymethylated β-cyclodextrin (BPCH-B) containing berberine (BBR) by a dissolution method. The BBR enhanced antimicrobial properties and maintained biocompatibility and drug delivery ability. The dressing was able to regulate drug release according to wound pH and temperature conditions, with a significant enhancement of drug release under alkaline conditions (pH 7.6, temperature 40 °C). The BPCH-B hydrogel demonstrated excellent wound healing, with healing area reaching 80% within 12 days. At the same time, BPCH-B hydrogel demonstrated strong antimicrobial activity in vitro, with drug release lasting for nearly 120 hours. BPCH-B hydrogel has the potential to be used as a dual-responsive antimicrobial dressing to aid in wound healing [10]. FGF-2 plays a key role in wound healing but has not been widely used in clinical wound therapy due to its instability and short half-life. Le et al. used carboxymethyl chitosan (CMCS)/hydroxyethylcellulose (HEC) hydrogel membranes, prepared by mixing and casting method, for the delivery and controlled release of FGF-2 to enhance its bioavailability in burn therapy. The hydrogel membranes demonstrated a porous structure, substantial swelling, promotion of cell proliferation, protection against FGF-2, and exhibited sustained release rates of more than 50% in aqueous solution. In a mouse burn model, FGF-2-containing hydrogel membranes significantly promoted early healing processes, including epidermal regeneration, granulation tissue formation, and angiogenesis, and slowed the development of hypertrophic scarring [11].

To address the challenges of diabetic trauma, Li et al. designed a versatile cryogel to cope with its complex microenvironment and healing process. The cryogel was prepared under sodium periodate oxidation with bio-friendly BC, gelatin and dopamine as the main components. BC enhances the mechanical stability of the gel and improves the backbone support effect and cross-linking degree. The gels presented excellent permeability and exudate management thanks to their interpenetrating porous structure. The gels exhibited efficient antimicrobial properties through I2 and sodium iodide produced by sodium periodate reduction. In addition, the cryogel induced macrophage polarization from M1 to M2, modulating the immune microenvironment in infectious diabetic trauma. The multifunctional cryogel effectively promoted collagen deposition and neoangiogenesis, thereby accelerating the healing of infected diabetic wounds [12] (Figure 2.2).

Morozova et al. successfully prepared cellulose-based hydrogels based on the Diels–Alder reaction using the covalent cross-linking reaction of two cellulose nanocrystals (CNCs) with furan and maleimide groups. And its gel properties at different temperatures were investigated. The results showed that its rheological properties could be modulated by adjusting the concentration and ratio of the components. The introduction of aldehyde groups into CNCs allowed the binding of physiologically active molecules containing primary amino groups via imine bonds. Benzocaine was used as a model drug for binding to CNC hydrogels, and the resulting drug-conjugated hydrogels exhibited good formulation stability at pH 7 and showed pH-sensitive release of benzocaine due to accelerated hydrolytic dissociation of the imine bond at pH < 7. This developed drug-conjugated hydrogel is promising as a wound dressing for local anesthesia [13].

Berglund et al. self-assembled nanofibrillar celluloses (CNFs) to form bio-based hydrogels by hydrogen bonding. They first isolated CNFs directly from wood, self-assembled hydrogels by vacuum-assisted filtration (VF), and compared them with commercial BC. The results showed that the self-assembled hydrogels from wood showed potential in wound dressings with comparable performance to

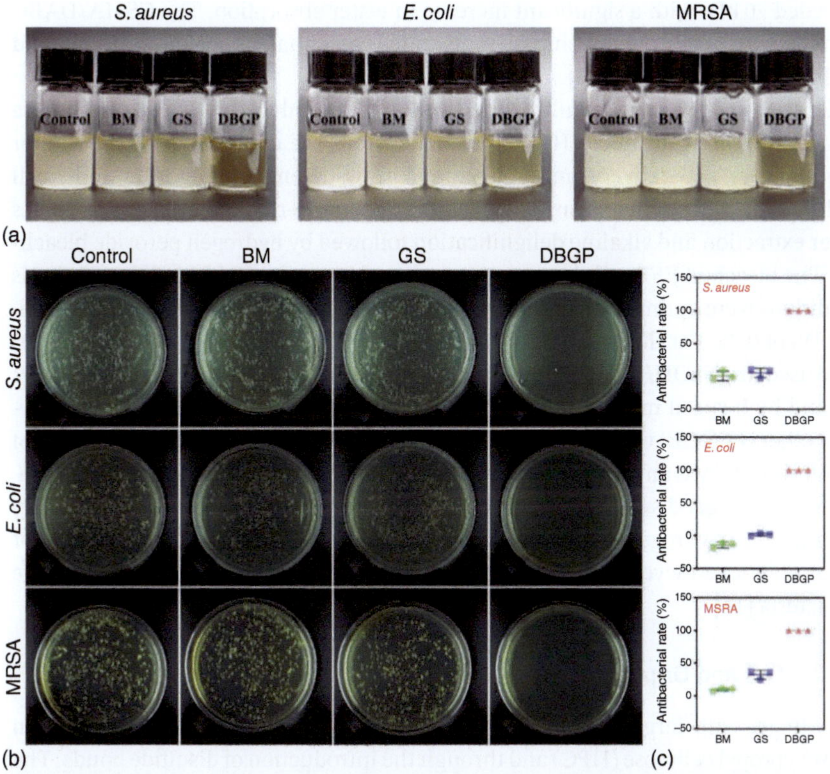

Figure 2.2 (a) After 12 hours of co-cultivation, the turbidity of the bacterial solution in the BM, GS, and DBGP hydrogel groups was significantly lower than that in the control group, indicating that the hydrogel had an antibacterial effect. (b) The colony counts of the hydrogel group were significantly lower than those of the control group, which directly proved the antibacterial activity of the hydrogels. (c) The three hydrogels showed good antibacterial activity against *Staphylococcus aureus*, *Escherichia coli*, and methicillin-resistant *S. aureus* (MRSA), and DBGP hydrogel was the most effective. Source: Ref. [12]/John Wiley & Sons/CC BY 4.0.

BC. Hydrogels prepared by vacuum-assisted filtration method performed well in mimicking BC performance. The choice of different separation raw materials and self-assembly techniques affects the network formation and structural stability of the hydrogels. Overall, this nanofibrillar cellulose hydrogel possesses scalable and sustainable properties, is designed to take full advantage of its intrinsic properties, exhibits strength comparable to human skin, and is suitable for wound dressing applications [14]. Deng et al. developed an all-natural self-healing hydrogel based on methacryloyl chitosan (CSMA) and dialdehyde-based bacterial cellulose (DABC). CSMA with water solubility and photocross-linking effect was obtained by modification of CS. DABC was oxidatively modified to introduce dialdehyde groups. The hydrogels were formed in a short time by the Schiff base group between the amino group of CSMA and the aldehyde group of DABC, which exhibited good self-healing properties. The hydrogel has a uniform porous structure with a uniform distribution of DABC nanofibers. The compressive strength of the hydrogel exceeded 20 kPa with a significant increase in water absorption. The CSMA/DABC hydrogel was cell-friendly and demonstrated potential applications in wound healing and tissue repair [15].

Hajidariyor et al. successfully synthesized a freeze-induced hydrogel of cellulose and carboxymethylcellulose (CMC) extracted from the cell wall of bromelain for drug delivery, cell delivery, and bone and skin tissue engineering to promote cell proliferation and regeneration. They obtained cellulose-rich materials by hot press water extraction and alkaline delignification followed by hydrogen peroxide bleaching. The bleached EFB cellulose was used as a substrate for CMC, and the synthesis conditions were optimized by fractional factorial design to obtain CMC with an optimal DS of 0.75. For the cold gel studies, cellulose, CMC, and β-cyclodextrin (β-CD) were used in a NaOH/urea system, and hydrogels were formed by spontaneous freezing and high-speed mixing. Epichlorohydrin (ECH) and polyethylene glycol diglycidyl ether (PEGDE) were used as cross-linking agents. By adjusting the ECH ratio of cellulose and CMC and adding different cross-linkers and matrices, the E-CMC-CEL hydrogel with good swelling properties and drug-loading release ability was finally obtained. The hydrogel proved to have excellent carrying and releasing properties for the antibiotic tetracycline and showed successful results in inhibiting a wide range of bacteria [16].

2.1.2 Cell and Organoid Culture

A hydrogel with large porosity has been fabricated by Sun et al. [17], based on hydroxypropyl cellulose (HPC) and through the introduction of disulfide bonds. The original physical and chemical properties of the hydrogel were maintained while giving it rapid cleavage and recovery performance. Compared with the traditional 2D cell culture, the time for this hydrogel to grow into a spherical shape of HepaRG can be shortened from 28 to 9 days. Multicellular carcinoma spheres are aggregates of cancer cells with strong correlation, which can simulate many characteristics of solid tumors on an in vitro platform using hydrogel as a carrier. However, traditional culture methods limit the separation of specific factors that affect the fate of cancer cells. To solve this problem, Li et al. [18] used poly(*N*-isopropylacrylamide) surface-functionalized cellulose nanocrystals to prepare supramolecular hydrogels with temperature response characteristics. When the temperature is higher than 37.8°, the material becomes a gel state, and when the temperature is lower than 37.8°, it can be converted into liquid to release the multicellular carcinoma spheres. On the basis of cell culture, Curvello et al. [19] further explored the possibility of cellulose-based hydrogels in preparing human organoids. They prepared an ionic hydrogel containing 0.1% cellulose nanofibers and functionalized it by grafting RGD peptides to induce the formation and growth of small intestinal organoids, creating a favorable environment (Figure 2.3).

Torres-Rendon et al. [20] noted that the internal pore structure of hydrogels plays an important role in the proliferation and differentiation of cells. In order to prepare a hydrogel with richer pores and a more stable structure, they reported a reverse printing strategy. They first prepared a sacrificial structure containing unstable cross-links, using cellulose and chitin nanofiber hydrogels. The sacrificial structure

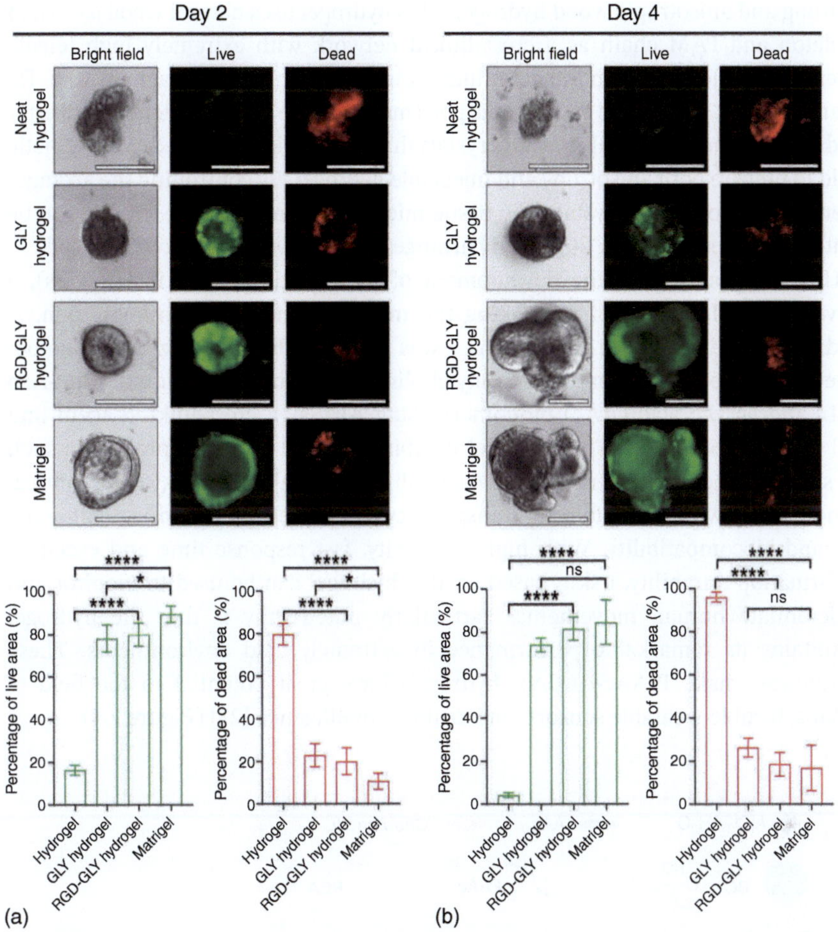

Figure 2.3 (a) Crypts cultured in pure gels showed large areas of dead cells, and cell survival was increased by the addition of GLY, but the morphology was different from that of the Matrigel control. The crypts cultured in RGD-GLY gel could form vesicle-like structures, and their development was similar to that of Matrigel. (b) After four days, RGD-GLY gel could induce the growth of crypt-tissue-forming buds, while without RGD only the volume increased. Source: Ref. [19]/John Wiley & Sons/CC BY 4.0.

is completely covered, and then the sacrificial structure is completely dissolved to finally prepare a cellulose hydrogel with abundant pore structure. He found human mesenchymal stem cells (hMSCs) can adhere well to the internal structure of the hydrogel, and the addition of calcium phosphate coating can further induce their osteogenic differentiation.

2.1.3 Tissue Regeneration

Inspired by the arrangement structure of muscle fibers, Kong et al. [21] combined tightly packed wood nanofibers with polyacrylamide (PAM) hydrogels to develop a strong and anisotropic wood hydrogel. This hydrogel uses natural wood as a rigid skeleton and PAM chain as a cross-linked network with extremely high tensile strength and flexibility, and its longitudinal tensile strength can reach 36 MPa. De France et al. [22] are based on isotropic poly(methacrylate) oligoethylene glycol ester hydrogels, by adding cellulose nanocrystals that can be highly aligned in a magnetic field to prepare both anisotropy and injectable hydrogel. By controlling the arrangement of cellulose nanocrystals, the tissue microenvironment of the muscle can be better simulated and then the orderly arrangement of myotubes can be induced.

He et al. focused on the development of biomimetic ionic skins (i-skins), a novel material capable of mimicking the multiple functions of human skin. A hydrogel-based i-skin (P[AAc-*co*-CA]$_x$) was successfully synthesized by one-pot free radical co-polymerization of vinylcholine-asparagine ionic liquid (Cho-Asn VIL) and acrylic acid (AAc) monomers using Al^{3+} as a cross-linking agent and BC as a reinforcement. The hydrogel exhibited several excellent properties such as skin-like modulus, ultra-high stretchability, self-healing ability, and electrical conductivity, while maintaining transparency, self-adhesion, anti-microbial activity, and biocompatibility. With high sensitivity, fast response time and excellent deformation durability, i-skin based on this hydrogel can be used to monitor and differentiate human movements. Particularly noteworthy is that the hydrogel maintains its remarkable performance in extremely cold environments. These advantages make P(AAc-*co*-CA)$_x$ hydrogels have great potential in the field of i-skins, flexible wearable sensors, and artificial intelligence [23] (Figure 2.4).

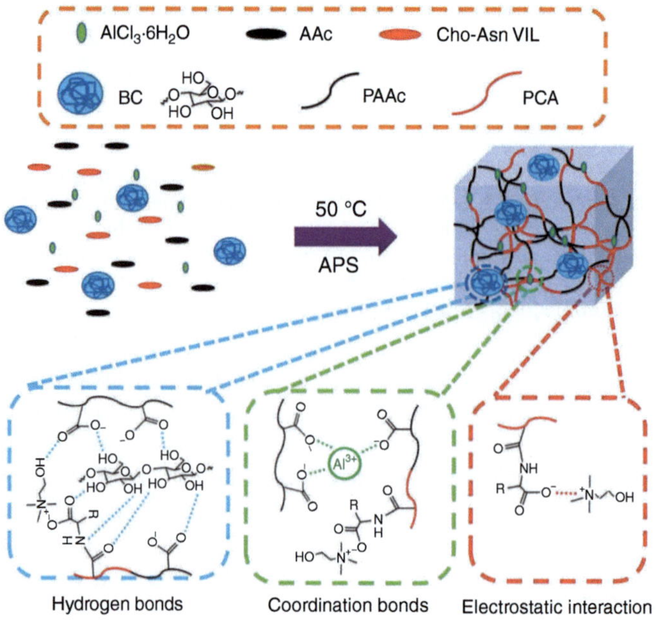

Figure 2.4 Schematic illustration for the preparation and chemical structure of P(AAc-*co*-CA)$_x$ hydrogels. Source: Ref. [23]/John Wiley & Sons.

Based on a "hard frame-soft permeation" design, Cai et al. successfully prepared an organic hydrogel with superior frost resistance and long-term water retention. The hydrogel was firmly attached to the tissue surface at low temperature (−80 °C) through "hydrogen-bonded clusters." The introduction of BC gives the hydrogel a high tensile strength, similar to that of skin. In vivo experiments have demonstrated that PRP-loaded hydrogels release multiple growth factors to accelerate wound healing. This strategy is expected to extend the lifetime of hydrogels in harsh medical environments [24].

Cellular perception and feedback of the material environment has been a hot research area in the field of tissue engineering. Ganguly et al. investigated the early osteoblastic differentiation of human bone-marrow-derived mesenchymal stem cells (hBMSCs) on very soft 3D hydrogels. Mechanotransduction, cytoskeletal distribution, antioxidant, calcium homeostasis, osteogenic marker gene expression, and cytokine secretion were observed by introducing nanofibrillar cellulose into methacrylamide gelatin hydrogel and applying short-duration pulsed pressure stimulation (PPS) at 5–20 kPa for seven days. RNA sequencing revealed differences in gene expression involving osteogenesis, bone mineralization, ion channels, and adhesion. This demonstrates a practical approach to drive osteogenic differentiation of hBMSCs on soft substrates that is highly scalable for clinical reconstruction [25].

There is an old Chinese proverb "Lotus root broken silk connected," and Guan et al. [26] conducted in-depth research on this phenomenon. They found that the high tenacity and high elongation of lotus root are related to its spiral microstructure. They first prepared the BC hydrogel, and then applied a tangential force on the cross-section to locally deform the cut hydrogel fibers. During the deformation process, the hydrogen bonds were broken by the tangential force. When the tangential force was removed, the hydrogen bonds between the nanofibers are changed, and the helical structure of the fibers is fixed. This hydrogel fiber with high strength and high toughness has great potential in surgical sutures. As the main load-bearing tissue of the spine, the intervertebral disc plays an important role in the movement of the human body, but it is precisely because of its strict weight load requirements that there is still a lack of suitable biological materials to replace the damaged disc. For this, Schmocker et al. [27] prepared an injectable polyethylene glycol dimethacrylate nanofiber cellulose hydrogel. They changed the molecular weight of the polymer and the concentration of nanocellulose to make it closer to the nature of the natural nucleus pulposus. Due to the injectable nature of this hydrogel, it can be used for minimally invasive injection for nucleus pulposus replacement therapy. In the bovine nucleus pulposus replacement experiment, the height of the intervertebral disc was restored from 65.6% to 99.0%, and the structural stability was maintained during 500,000 loading cycles.

Cartilage tissue engineering is a novel therapeutic modality used for the treatment of osteochondral defects including injuries to the osteoarticular surfaces. Chayanun et al. solved the problem of low stiffness by introducing nanofibrillar cellulose (NFC) into acetaldehyde-gelatin (ADA-GEL) hydrogels. The NFC-reinforced ADA-GEL (AG-N) composite hydrogels exhibited significantly increased stiffness.

The results of the study showed that AG-N hydrogels maintained stiffness at room temperature and had good chondroinductive activity. The AG-N structure remained intact for 28 days despite NFC affecting the water absorption properties of the hydrogel. Temperature stability remains an issue, and attention needs to be paid to cross-linking the gelatin to ensure adequate stiffness. AG-N hydrogels had no negative effect on cellular responses and their chondrogenic differentiation capacity was remarkable. We also present possible applications for the use of AG-N hydrogels as osteochondral plugs, providing promising biomaterials for the treatment of osteochondral defects [28].

The weak mechanical properties of traditional hydrogels limit their application as flexible sensors. Improving the mechanical properties and electrical conductivity of hydrogels is the key to realize the application of flexible electronics. Zhong et al. used the synergistic effect of pre-strain and ionic cross-linking to construct cellulose-based dual-network oriented hydrogels and regulate their internal structure. The pre-strain enhances the chemical cross-linking network and the ionic cross-linking generates the physical cross-linking network, which synergize with each other to improve the mechanical strength. At the same time, the pre-strain-induced alignment produced anisotropy in the hydrogels. The hydrogels prepared by this method showed a threefold increase in tensile strength, a 1.3-fold increase in light transmittance, and improved thermal stability and electrical conductivity. The dual-network oriented hydrogels can be used as flexible substrates for the preparation of highly sensitive strain sensors for precise monitoring of various human body movements [29]. Sun et al. developed a novel cellulose-based composite hydrogel for the preparation of stretchable and self-powered electronic devices. The material was prepared by single-step polymerization of hydroxyethyl cellulose macromolecule monomer (HECM) with acrylate and hydroxyl groups and acrylic monomer to achieve both chemical and physical cross-linking, and the introduction of HECM significantly enhanced the cross-linking density and improved the mechanical properties of the material. At the same time, the in situ polymerization of conductive polymers achieves excellent electrical properties. The material shows excellent self-enhancement effect and can be used to prepare high sensitivity strain sensors. As electrode materials, stretchable nanogenerators can be prepared for energy harvesting and self-powered sensing of human movement. This green preparation and multifunctional cellulose hydrogel provide a new strategy for sustainable and environmentally friendly electronic device innovation [30]. Based on the synergistic effect of hydrogen bonding and bimetallic coordination bonding of bialdehyde cellulose and carboxyl groups with Fe^{3+}, Han et al. prepared cellulose-based hydrogels with dual shape memory effect, high tensile strain (up to 600%), good self-recovery, and anti-fatigue properties, which have a great potential for application in the field of human motion monitoring [31]. Liu et al. prepared hydrogels containing CNF, polypyrrole (PPy), and glycerol using a one-pot method. The addition of PPy improved the electrical conductivity (~0.034 S/m), and the ionic conductivity at −18 °C was comparable to room temperature. The CNF and PPy polymer chains in the hydrogel structure formed strong hydrogen bonds, which significantly improved the mechanical strength and viscoelasticity. With a high strain gain factor (GF of 2.84), this novel hydrogel exhibits excellent stability and is suitable for human activity monitoring [32].

Danmatam et al. successfully prepared a hydrogel with electroactive and self-healing capabilities based on several components, namely carboxymethyl cellulose (CMC), polythiophene (PTh), and acid-hydrolyzed cellulose (AHC). During the preparation process, the AHC particles were effectively dispersed in the CMC matrix, which improved the thermal stability of the hydrogel. By applying an electric field, the hydrogel showed an electrically responsive property of bending toward the cathode electrode. They found that the AHC content was negatively correlated with the electroactivity of the hydrogels. In terms of self-healing, the hydrogel with 2 wt% AHC added exhibited the highest self-healing efficiency in terms of tensile strength, elongation at break, and bending angle, and after 24 hours of self-healing, the performance of the CMC/PTh/AHC2 hydrogel was close to the original level. This hydrogel has potential applications as an actuator or artificial muscle under electrical stimulation [33].

Based on HPC with hydrophilicity liquid crystal and thermochromism and 2-acrylamide-2-methylpropene sulfonic acid (PAMPS) with polypropylene sulfonic group, Shi et al. developed a new type of anisotropic hydrogel. Through mechanical cutting, HPC molecules are oriented, and this arrangement is locked through the hydrogel network to form a large area of unidirectional structure. The anisotropic hydrogel shows uniform interference color between cross-polarizers, which can be used as a multi-functional sensor and is responsive to temperature, pressure, and tension. In particular, the directional HPC/PAMPS hydrogel has the ability to respond as a temperature sensor in a wide temperature range, and because of the constraints of the hydrogel network, its transmission brightness loss is small when the interference color changes. This study provides a new idea for the application of anisotropic hydrogels in the field of sensors [34].

2.2 Cellulose-Based Electrospinning for Biomedical Application

2.2.1 Drug Delivery

As the electrospinning solution can be adjusted according to requirements, for example, the electrospinning fiber membranes with different pharmacological effects can be prepared by blending or covalent bond cross-linking. This makes cellulose electrospun fibrous membranes have numerous applications in the field of drug delivery.

Human immunodeficiency virus (HIV) is a virus that has not yet been overcome by medicine. It endangers the health of millions of people worldwide. Currently, there is no specific medicine for the treatment of human immunodeficiency virus, so most of the current work is focused on how to effectively control the spread of AIDS. Sexual transmission is a major form of AIDS. Based on this, Huang et al. [35] used the different acid–base microenvironments of vagina and semen to

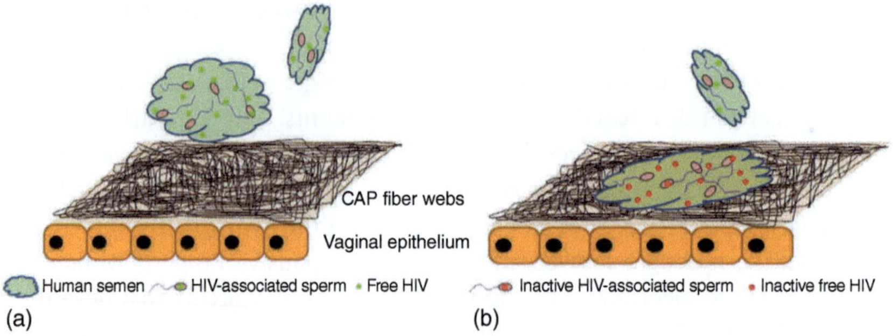

Figure 2.5 (a) Vaginal tissues were covered with electrospun carboxylic acid cellulose (CAP) fiber mesh containing antiviral drugs. (b) A vaginal drug delivery platform was constructed using electrospun CAP fibrous mesh containing antiviral drugs, and the surface of the vaginal tissue was still covered by the CAP fibrous mesh after contact with HIV-containing human semen. Source: Ref. [35]/Elsevier.

prepare pH-responsive electrospinning biomaterials. They loaded antiviral drugs in electrospun fibers made of cellulose acetate (CA) phthalate. Fiber can keep the fiber intact in the acidic environment of the vagina. When alkaline semen enters the vagina, it will quickly dissolve and release antiviral drugs to effectively block the spread of AIDS (Figure 2.5).

Allafchian et al. [36] used the covalent connection between the carboxyl group contained in carboxymethyl cellulose and the amino group contained in the loaded drug to significantly extend the biological half-life of the drug, and it can be further adjusted by adding a cross-linking agent. Due to the low bioavailability of paclitaxel, and the large amount of chemotherapeutic drugs are extremely harmful to the human body, improving the anti-tumor activity of paclitaxel is very important for cancer treatment. Liu et al. [37] prepared an electrospun fiber membrane for enhancing the anticancer effect of paclitaxel by co-electrospinning paclitaxel, rectorite, and CA solution. The addition of rectorite reduces the overall fiber diameter on the one hand and controls the release rate of paclitaxel on the other hand. Although the release rate of paclitaxel was reduced, its cytotoxic effect on human gastric adenocarcinoma SGC7901 cells growing on the nanofiber mat was enhanced.

Controlled drug release by microcurrent stimulation is a novel area of research, and Lago et al. investigated controlled drug release by electrical stimulation by using electrospun CA membranes and CA membranes containing ibuprofen (IBU). This type of fiber membrane can control drug release by very low potentials. Application of a negative potential of −0.3 V showed retention of ibuprofen, while positive potentials of +0.3, +0.5, or + 0.8 V enhanced drug release. The feasibility of this system for dermal applications was verified by fabricating small patches and validating the "on/off" release pattern of ibuprofen in the membrane. This work tests different systems for achieving electrically controlled drug release from the skin, providing a valuable contribution with potential personalized medicine applications [38].

Amani et al. prepared a pH-sensitive bilayer electrospun nanofiber material as a novel burn wound dressing using electrospinning technology. The top layer of the material consisted of ethyl cellulose (EC), and the bottom layer consisted of ethyl cellulose/Eudragit S-100 (EC/ES-100). The bilayer nanofiber film exhibited different dissolution rates and drug release properties at different pH conditions. MTT test results showed that the fiber film had a good safety profile for cells. The pH-sensitive bilayer electrospun nanofiber film showed superior performance in promoting burn wound healing and was more rapid and effective than other bilayer films [39].

2.2.2 Antibacterial

Electrospun membranes made of cellulose and its derivatives have excellent biocompatibility, biodegradability, and have the potential to enhance the cell interaction between the scaffold and fibroblasts, and are widely used in drug delivery to promote wound healing and other areas. At present, the improvement of electrospinning membrane performance is mainly concentrated in the following two aspects: one is to optimize the components of the electrospinning solution to give the electrospinning membrane different biological functions; the second is to change the electrospinning nozzle structure and improve electrospinning methods. The electrospun membrane with multi-layer structure is prepared by means of postcross-linking or layer-by-layer stacking. Chen et al. [40] achieved sustained drug release by preparing a porous CA membrane loaded with thymol. Its slow-release ability is mainly achieved by trapping air in the porous structure to increase the hydrophobicity of the fiber membrane. In addition, the larger the specific surface area of the porous fibrous membrane loaded with thymol, the higher the drug utilization rate and the better cell compatibility. Bacteriostatic experiments show that this porous electrospun membrane has a significant antibacterial effect on *Staphylococcus* and *E. coli*. In terms of bacteriostasis, unlike the previous direct use of electrospinning to load bacteriostatic drugs, Wang et al. [41] prepared a photodynamic antibacterial film by embedding the photosensitizer protoporphyrin IX in situ. Under light conditions, protoporphyrin IX can produce singlet oxygen and kill bacteria. Unlike ordinary skin damage, the high blood sugar of diabetic patients can cause various complications and cause their wounds to heal very slowly. Liu et al. [42] found through research that a blended electrospun nanofiber membrane prepared from CA and zein can reduce the expression of interleukin-10, by loading sabacol (up-regulating interleukin 10) that has a synergistic effect with the fibrous membrane, which can promote the proliferation of keratinocytes in diabetic mouse wounds and promote wound healing (Figure 2.6).

Silver nanoparticles have excellent antibacterial properties and are widely used in wound accessories. However, long-term skin contact with silver will cause silver deposition and increase the chance of cancer. Therefore, improving the stability of silver-containing biomaterials is of great significance for reducing skin toxicity and extending the antibacterial time of the material. Jatoi et al. [43, 44] reported two ways to improve the stability of silver nanoparticles. A method of anchoring silver nanoparticles on carbon nanotubes to enhance material stability. They first produced carbon nanotubes/silver nanocomposites by oxidizing carbon nanotubes to generate carboxyl and hydroxyl groups to adsorb silver ions, and then embed the composites in a CA matrix to prepare electrospun fiber membranes. The other

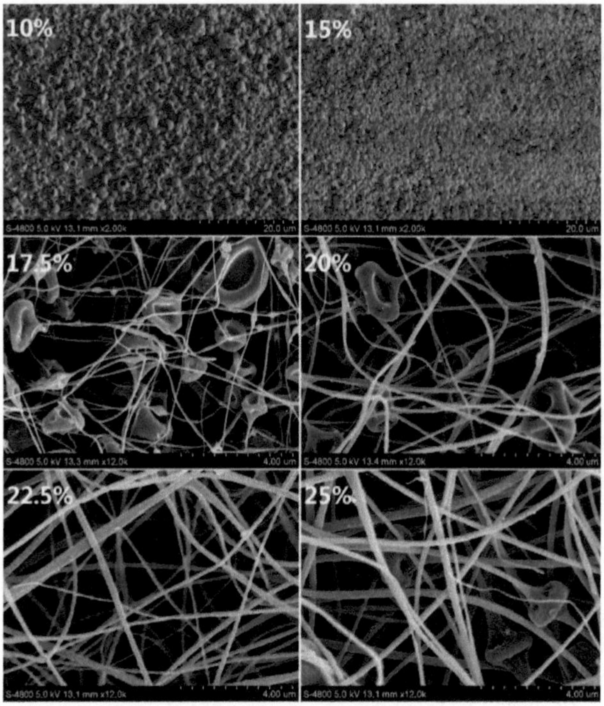

Figure 2.6 Effect of cellulose acetate concentration (wt%) on fiber morphology. Source: Ref. [42]/Elsevier.

is to use the strong adhesion of dopamine to adsorb silver on the surface of ZnO, then use its metal reduction effect to turn silver ions into nanoparticles, and finally use electrospinning technology to prepare fiber membranes. These methods largely avoid the direct contact of silver with the skin, and at the same time, they will not affect the ability of silver nanoparticles to generate ROS, and improve safety while ensuring the antibacterial ability of the material.

2.2.3 Tissue Regeneration

Due to their unique physical and chemical properties, electrospun fiber membranes of cellulose derivatives are also popular in the field of tissue engineering. In many cases, it is difficult for a single-structure nanofiber membrane to meet the diverse needs of promoting wound healing, so the innovation of the electrospun fiber membrane structure is particularly important. Yang et al. [45] first introduced the Janus nanostructure concept to the preparation of electrospun membranes. They developed an eccentric spinneret to eliminate the repulsive force between two side-by-side spinning solutions. They carried silver nanoparticles and ciprofloxacin together in Janus fiber membranes made of polyvinylpyrrolidone and ethyl cellulose. On the one hand, the new Janus wound dressing helps to stop bleeding and increase the absorption rate of wound exudates. On the other hand, different types of active ingredients can be loaded into the Janus structure at the same time, so that both sides are in contact with the environment and produce a synergistic effect. Ramanathan et al. [46] combined electrospinning technology and freeze-drying technology to prepare a cellulose double-layer electrospun membrane. The top layer is composed of hybrid CA nanofibers containing bioactive latex or ciprofloxacin. The bottom layer is composed of a highly interconnected collagen matrix. While having an effective antibacterial effect, the porous structure of the bottom layer is conducive to the absorption of wound exudate and cell adhesion.

Nada et al. [47] prepared a new type of wound dressing by laminating electrospun fibers with different functions. The first layer is made of synthetic hemostatic agent chitosan iodoacetamide loaded into electrospun polyvinyl alcohol fiber, which plays a role of hemostasis; the second layer is gelatin nanometer cross-linked by glyoxal fiber wraps up to 50% (wt%) of capsaicin to make an analgesic layer; the third layer is polyvinyl alcohol electrospun fiber, loaded with 2% (wt%) gentamicin, made Get an antibacterial layer. The multilayer film of this "hamburger" structure can effectively accelerate wound healing on the basis of maintaining good biocompatibility. Ruhela et al. [48] used CA to prepare a cellulose electrospun membrane with "semipermeable membrane" properties and used it in islet cell culture. In vivo experimental results confirmed that the hydrophobicity and surface morphology of the electrospun membrane hindered cell attachment, which helped maintain the three-dimensional circular morphology of pancreatic islet cells. More importantly, the pore size obtained due to the densely packed structure of nanofibers is 0.3–0.6 μm, which will be able to restrict immune cells from entering the electrospun membrane to kill islet cells, but allows free movement of molecules such as insulin and glucose.

Abdulmalik et al. improved the therapeutic efficacy of tendon injuries by using enhanced nanofiber matrix delivery of growth factor (Ex-4). They used a hybrid polycaprolactone (PCL) and CA method combined with electrospinning to prepare micro-nanofiber membranes with certain structures. Ex-4 was encapsulated in halloysite nanotubes (HNTs), which successfully reduced burst release. It was demonstrated by in vitro cell culture experiments that Ex-4 nanofiber membranes promoted the proliferation of hMSCs and the expression of tendon-related genes, and reduced the formation of fibrocartilage. This suggests that Ex-4 nanofibrous matrix has potential therapeutic applications in tendon tissue engineering and provides a useful strategy for future clinical applications [49].

In order to simulate the fibril structure of natural ECM, Luo et al. [50] hybridized BC nanofibers and CA submicron fibers for the first time to prepare an intertransmissive nano-submicron fiber scaffold. BC nanofibers (42 nm in diameter) can penetrate into the submicron fiber CA scaffold with a diameter of 820 nm, and the porosity of the overall scaffold is above 90%. And the combination of CA submicron fibers and BC nanofibers significantly enhances the overall mechanical properties of the fiber scaffold. The results show that the BC/CA multi-layer scaffold is beneficial to cell migration and proliferation, while effectively increasing the expression of osteogenic genes and protein. It is difficult to prepare a tough three-dimensional scaffold from a single cellulose through electrospinning technology. Lee et al.

[51] used calcium hydroxide one-step electrospinning method to convert the three-dimensional CA/lactic acid nanofiber network into cellulose/lactic acid. The CA and lactic acid components in the original network are simultaneously converted into cellulose and calcium lactate components. This method enhances the overall structural strength (from 0.112 to 1.382 MPa) while improving the effect of inducing calcium deposition in the body. Patel et al. [52] deeply studied the effect of high aspect ratio cellulose nanocrystals on the biocompatibility and osteogenic ability of polylactic acid. They found that the mechanical properties of the nanofiber scaffold prepared by electrospinning after blending the two have been significantly improved, which may be due to the strong interaction between the polymer chain and the carbon nanotubes. In addition, compared with the pure polymer composite scaffold, the surface mineralization degree is better, and the expression of osteogenic gene markers is significantly increased, confirming its enhanced osteogenic ability.

Mohammadalipour et al. prepared a novel biomimetic ECM scaffold by mixing polyhydroxybutyric acid (PHB) with lignin and CNF. The results showed that this biomimetic scaffold has more symmetrical round fibers, higher surface roughness (from 326 to 389 nm), better hydrophilicity (from 120° to 60°), and more reasonable biodegradation properties compared to pure PHB scaffolds. These changes not only improved mechanical properties (toughness factor from 300 to 1100), but also improved cell behavior (cell survival from 60% to 100%), bioactivity (Ca/P ratio from 0.77 to 1.67), and higher levels of Alizarin Red staining and ALPase secretion. Safety and efficacy in a rat Achilles tendon defect model were verified by histology, marker expression, gait analysis, and mechanical testing, which demonstrated the significant advantages of PLC scaffolds over fibrin scaffolds alone in promoting tendon healing and reducing fibrocartilage formation [53].

Based on a high degree of simulation of the physicochemical environment of the extracellular matrix, Gonçalves et al. constructed organoid models of the intestine, skin, and lung using electrospun membranes of PCL and cellulose derivatives (CA, CAP, EC, and MC). The various scaffolds showed biocompatibility for most cell types, with the PCL–cellulose derivative mixture promising as a scaffold material suitable for in vitro epithelial tissue modeling and toxicity screening. In vitro experiments demonstrated that intestinal, skin, and lung cells exhibited different behaviors on these scaffolds, opening up potential applications of these biomaterials for in vitro tissue and organ modeling and drug discovery [54] (Figure 2.7).

Szewczyk et al. explored the potential of CA electrospun fibers for bone tissue regeneration. The piezoelectric properties of electrospun CA fibers were demonstrated for the first time by high-voltage switching spectroscopic piezoresponse force microscopy tests, and surface potential studies were performed by Kelvin probe force microscopy and zeta potential measurements. CA fibers exhibited high piezoelectric coefficients (6.68 ± 1.70 pm/V), high surface potentials (718 mV), and zeta potentials (−12.2 mV), which are highly similar to the natural electrical environment conducive to bone cell attachment and growth and could induce osteoblasts to attach and grow. The natural electrical environment that favors osteoblast attachment and growth is highly similar to that of osteoclasts and can lead to the formation of a large collagen three-dimensional network by osteoblasts [55].

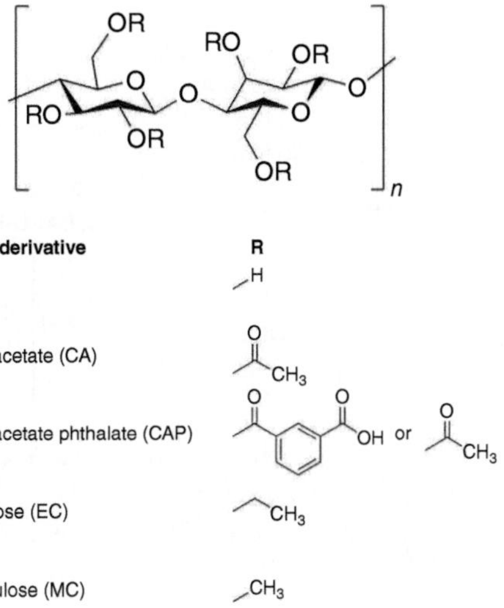

Figure 2.7 Chemical structure of cellulose and its derivatives. Source: Ref. [54]/John Wiley & Sons.

Ganguly et al. investigated the role of short-duration pulsatile pressure mechanical stimulation on osteogenic differentiation of human bone marrow mesenchymal stem cells on soft hydrogel scaffolds. Pulsating pressure of 5–20 kPa (mean: 15.78 ± 4.5 kPa) was applied to stem cells cultured on methyl methacrylate gelatin hydrogel containing cellulose nanocrystals (stiffness only 70 ± 0.00049 Pa) for seven days. The results showed that this stimulation induced a series of cellular responses that promoted the differentiation of stem cells into osteoblasts, including cell adhesion, cytoskeletal remodeling, antioxidant, calcium homeostasis regulation, up-regulation of bone-specific gene expression, and altered cytokine secretion. This promotion may be realized through mechanotransduction channels such as Piezo 1. In addition, genomic analyses showed that the stimulation led to alterations in the expression of several genes related to bone formation and mineralization. In conclusion, short-term pulsatile pressure stimulation can significantly promote the differentiation of stem cells to osteoblasts on soft hydrogel within a short period of time, which provides a simple and efficient new strategy for bone tissue engineering [25].

2.3 Cellulose-Based 3D Printing for Biomedical Application

2.3.1 Improvement of Bio-Ink

Nano-cellulose-based nanocomposite hydrogels can be used to prepare various biological materials through 3D printing technology to meet the different needs of the biomedical field. However, the addition of conventional cross-linking agents will cause an unavoidable drop in the mechanical properties of the hydrogel. Based on this, Wang et al. [56] used a photoactive bis(acyl)phosphine oxide derivative to modify the surface of cellulose nanocrystals. They reported a new BAPO derivative that can be directly attached to cellulose nanocrystals without any pretreatment. This

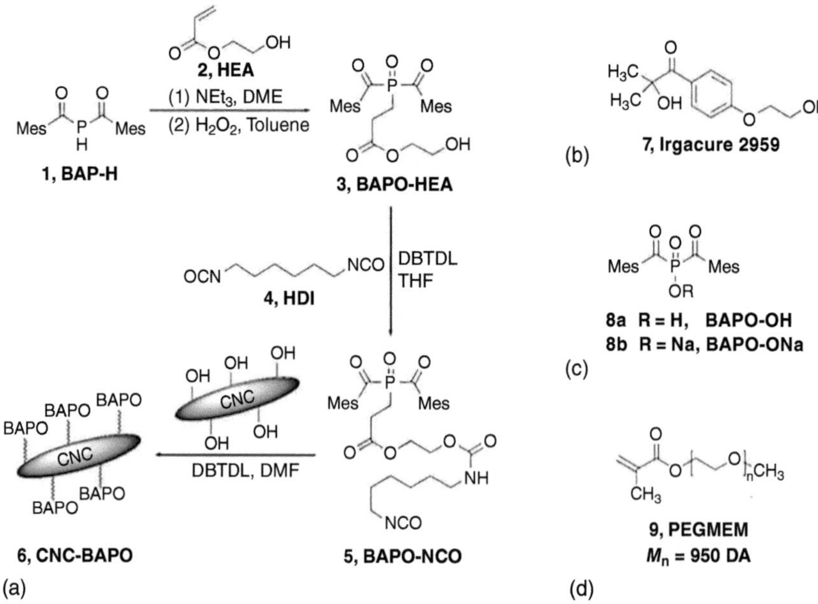

Figure 2.8 (a) Synthesis of BAPO derivatives and functionalization of cellulose nanocrystals, (b) BAPO-OH and BAPO-ONa, (c) and the mono-functional methacrylate PEGMEM, (d) Mes = mesityl = 2,4,6-$Me_3C_6H_2$. Source: Ref. [56]/John Wiley & Sons.

photoactive nanomaterial can convert the traditional monofunctional monomer into a polymer network without any additional cross-linking agent. The hydrogel prepared based on this material shows excellent swelling ability and improved mechanical properties (Figure 2.8).

Guo et al. developed a novel printable light-curing hydrogel material for efficient and rapid 3D printing of various shape structures. The mechanical properties of the material were significantly improved by introducing CNF as the reinforcing phase using both chemical and physical cross-linking strategies. The excellent toughness and compression resilience of the materials can be used to prepare flexible strain sensors for human motion monitoring, which can output stable electrical signals. The introduction of micelles not only promotes the light-curing reaction, but also provides a large number of sacrificial bonds to improve the toughness of the material. The printed material exhibits super toughness and resilience, which can be used for flexible and self-powered human motion detection. The controlled light-curing printing technique provides a new method for the rapid preparation of personalized hydrogel devices [57].

Bioprinting of microstructures with high precision has been a hot research area. Chen et al. applied concentric electrospinning to prepare nucleus-sheath microtubules to mimic the microstructure of capillaries, and then embedded them into carboxymethylcellulose/sodium alginate hydrogel for bioprinting in order to construct vascularized soft-tissue scaffolds with microchannels. It was found that microtubule concentration and length significantly affected the printed filament bundle size and merging area. By optimizing the microtubule concentration, it was possible to improve the printing fidelity, increase the microtubule density, and promote cell attachment. The experimental results showed that the microtubules prepared by electrospinning were not only compatible with human umbilical vein endothelial cells, but also provided microscopic topological signals for cell proliferation and morphogenesis in three-dimensional space. This study provides a potential solution for bioprinting vascularized tissue-engineered scaffolds for further tissue and organ modeling and preclinical drug development [58].

2.3.2 Bacteria and Cell Culture

3D printing bio-inks prepared from cellulose and its derivatives have become commonplace. Introducing bacteria into 3D printing inks to produce hydrogels spontaneously is a new idea. Schaffner et al. [59] used 3D printing technology to locate *Trichoderma* in different compartments. Unlike simply culturing bacteria on the surface of the hydrogel, the bacteria are uniformly combined in the entire printed hydrogel. In addition, multiple bacterial strains can potentially be located at specific locations in a complex architecture to study quorum sensing, bacterial growth, and migration. Compared with bacteria that only live on the surface, buried bacteria are also more resistant to external adverse effects (such as toxic ingredients). Fixing *Trichoderma* in a pre-designed 3D matrix can form a BC scaffold in situ on a non-planar surface, which has great application potential in the biomedical field. The metabolism of *Gluconacetobacter xylinus* is activated by oxygen, and the sufficient circulation of oxygen largely determines whether the bacteria wrapped in 3D printing ink can better exert their biological effects. Shin et al. [60] reported on the solid matrix-assisted 3D printing culture medium surface and the preparation method of BC hydrogel. They use polytetrafluoroethylene (PTFE) particles as a printing matrix and a hydrogel ink containing culture media, bacteria, and CNF. Due to the permeability of oxygen through the layer of polytetrafluoroethylene particles, biologically active bacteria produce BC hydrogel on the surface of the culture medium. The resulting tubular BC structure can be used in the preparation of artificial blood vessels and engineered blood vessel tissue scaffolds.

It is difficult for a single-component bio-ink to meet the dual requirements of providing mechanical support and cell culture at the same time. Fan et al. [61] used two different hydrogel inks sharing the same photocross-linking chemistry to prepare a seamless structural integration. They used cellulose nanocrystals and GelMA/HAMA as structural components, and GelMA/HAMA as cell gel, and used the mouse chondrocyte cell line ATDC5 to cross them. The coupling conditions have been optimized. The seamless scaffold prepared by 3D printing technology will basically not affect the activity of mouse chondrocyte line ATDC5, and can still maintain the integrity of the structure after 10 cycles of 20% strain, and can be used for cartilage repair. Cheng et al. [62] reported a new matrix-assisted sacrifice 3D printing strategy. They deposited the hydrophobic escape ink directly in the BC hydrogel matrix. After the BC hydrogel dries to form a paper-like film, the escape ink is removed to form hollow, pourable microchannels. Then HUVECs were planted into microchannels to form vascular structures, and MCF-7 cells were inoculated into the matrix surrounding the hydrogel to establish a three-dimensional paper vascularized breast tumor model for in vitro anti-cancer drug screening provides a feasible solution.

Three-dimensional printing technology offers new possibilities for constructing cellular microenvironments. However, existing 3D-printed cell culture scaffolds have the problem of limited structural and functional customization. In order to break through this limitation, Rad et al. developed a novel superparamagnetic 3D-printed cell culture scaffold by adopting the origami design concept. The scaffold is made of iron oxide nanoparticles and CA to achieve external magnetic control and structural remodeling. Unlike conventional scaffolds, which are fixed and unchanging, the scaffold can be folded to regulate the gravity environment and migration channels of cells, thus realizing dynamic control of the cellular microenvironment. By taking advantage of this feature, the researchers realized three-dimensional culture and diffusion of yeast and mouse fibroblasts. The cells were able to undergo gravitational control and direct migration with the folding of the scaffolds, and realized spatio-temporal dynamics of cell proliferation. This study opens up a new way of cell culture and supports the multidirectional interaction of cells with the microenvironment and other cells, which is expected to be a powerful tool for monitoring cell response, drug screening, and cell therapy [63] (Figure 2.9).

CA is a natural biodegradable material with a wide range of applications in tissue engineering and plant science research, but its poor mechanical properties limit the applications. Three-dimensional printing technology offers new possibilities for constructing CA materials, but traditional methods have limited results in improving mechanical properties only by increasing the concentration of the starting material. In order to achieve accurate control of the properties of printed CA, Slejko et al. implemented a continuous printing method by modifying a commercial 3D printer. The study systematically examined the effects of printing parameters and

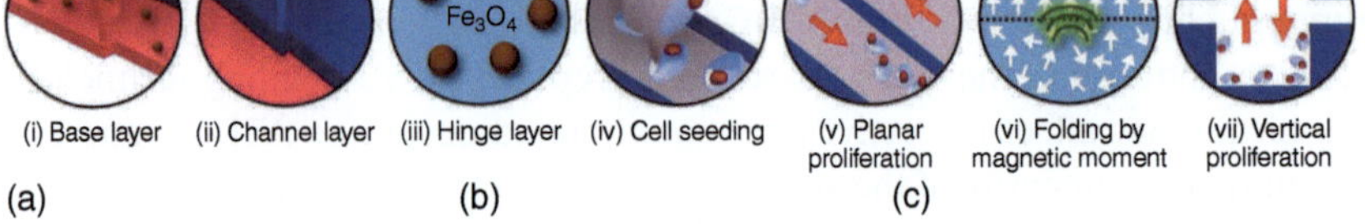

Figure 2.9 (a) Step-by-step 3D printing of a magnetic scaffold including substrate, channel and hinge layers, and incorporation of cells inside the channel. (b) Before the scaffold was folded, cells grew in the plane of the channel. (c) After the scaffold was folded, cells grew in the vertical direction of the channel under the influence of the magnetic field. Source: Ref. [63]/John Wiley & Sons.

ink concentration on material density, mechanical strength, and water absorption. The results showed that the appropriate combination of parameters could effectively control the final material properties; lower-density samples absorbed water faster and had higher water content; and the diffusion of water within the material conformed to Fick's law. This study establishes a method to precisely regulate the properties of 3D printed CA, which has obvious advantages over the traditional method [64].

Diatoms are classical bioindicators for assessing water pollution, but traditional methods are time-consuming and require specialized equipment. In order to realize rapid and simple water quality monitoring, Boons et al. prepared a hydrogel bioindicator containing bioactive diatoms using diatoms, cellulose nanocrystals, and sodium alginate as raw materials by using three-dimensional printing technology. Cellulose nanocrystals played a role in improving the rheological properties of the ink and providing mechanical support. The optimized hydrogel printing material could effectively fix the diatoms and make them survive and proliferate. Different diatoms have different sensitivities to pollutants, and the water quality can be quickly determined by comparing the changes in the number of diatoms after one to two weeks of incubation. This study utilizes the metabolism of microorganisms to prepare a simple bioindicator of water quality, which provides a new idea based on three-dimensional printing and microorganisms in the field of medicine [65].

Cellulose has the potential to be an excellent insulating material, but cellulose aerogels prepared by conventional methods have limited properties. To break through this limitation, Nagel et al. prepared functional cellulose-based aerogels by direct writing, taking advantage of three-dimensional printing technology. The formulation of cellulose hydrogel was optimized to meet the printing requirements. The results show that the material is anisotropic due to the shear force during the printing process; after drying, the thermal conductivity and mechanical strength of the material are better than those of the traditional aerogel. This study realizes the precise control of the properties and structure of cellulose aerogel, and provides a new strategy for the simple preparation of high-performance cellulose insulating materials. In addition, the natural degradability of cellulose also gives the material potential for biomedical applications, such as the preparation of gradual degradation of implantable materials or drug delivery systems, to further optimize biocompatibility and degradation kinetics [66].

References

1 Moon, R.J., Martini, A., Nairn, J. et al. (2011). Cellulose nanomaterials review: structure, properties and nanocomposites. *Chem. Soc. Rev.* 40: 3941–3994.

2 Lakhundi, S., Siddiqui, R., and Khan, N.A. (2015). Cellulose degradation: a therapeutic strategy in the improved treatment of infections. *Parasit. Vectors* 8: 23.

3 Wang, Y.F., Cooke, M.J., Morshead, C.M., and Shoichet, M.S. (2012). Hydrogel delivery of erythropoietin to the brain for endogenous stem cell stimulation after stroke injury. *Biomaterials* 33: 2681–2692.

4 Appel, E.A., Loh, X.J., Jones, S.T. et al. (2012). Sustained release of proteins from high water content supramolecular polymer hydrogels. *Biomaterials* 33: 4646–4652.

5 Zhang, K.Y., Wu, D., Chang, L.M. et al. (2023). Cellulose based self-healing hydrogel through boronic ester connections for wound healing and antitumor applications. *Int. J. Biol. Macromol.* 230: 123294.

6 Zhao, X.Q., Shi, Y.C., Niu, S.F. et al. (2024). Enhancing wound healing and bactericidal efficacy: a hydrogel membrane of bacterial cellulose and Sanxan gel for accelerating the healing of infected wounds. *Adv. Healthc. Mater.* e2303216. https://doi.org/10.1002/adhm.202303216.

7 Zhang, J.Y., Zhang, G.N., Wang, Y. et al. (2023). L-Arginine carboxymethyl cellulose hydrogel releasing nitric oxide to improve wound healing. *Eur. Polym. J.* 189: 111940.

8 Zhong, W., Hu, R.J., Zhou, S. et al. (2023). Spatiotemporally responsive hydrogel dressing with self-adaptive antibacterial activity and cell compatibility for wound sealing and healing. *Adv. Healthc. Mater.* 12: 2203241.

9 Liu, X.A., Zhang, Y.J., Liu, Y.J. et al. (2023). Injectable, self-healable and antibacterial multi-responsive tunicate cellulose nanocrystals strengthened supramolecular hydrogels for wound dressings. *Int. J. Biol. Macromol.* 240: 124365.

10 Zhang, Y.T., Gao, X., Tang, X.N. et al. (2023). A dual pH- and temperature-responsive hydrogel produced in situ crosslinking of cyclodextrin-cellulose for wound healing. *Int. J. Biol. Macromol.* 253: 126693.

11 Le, K.T., Nguyen, C.T., Lac, T.D. et al. (2023). Facilely preparing carboxymethyl chitosan/hydroxyethyl cellulose hydrogel films for protective and sustained release of fibroblast growth factor 2 to accelerate dermal tissue repair. *J. Drug Deliv. Sci. Technol.* 82: 104318.

12 Li, Y., Yang, Z.F., Sun, Q. et al. (2023). Biocompatible cryogel with good breathability, exudate management, antibacterial and immunomodulatory properties for infected diabetic wound healing. *Adv. Sci.* 10 (31): 2304243.

13 Morozova, S.M. and Korzhikova-Vlakh, E.G. (2023). Fibrillar hydrogel based on cellulose nanocrystals crosslinked via Diels–Alder reaction: preparation and pH-sensitive release of benzocaine. *Polymers (Basel)* 15: 4689.

14 Berglund, L., Squinca, P., Bas, Y. et al. (2023). Self-assembly of nanocellulose hydrogels mimicking bacterial cellulose for wound dressing applications. *Biomacromolecules* 24: 2264–2277.

15 Deng, L.L., Ou, K.K., Shen, J.X. et al. (2023). Double cross-linked chitosan/bacterial cellulose dressing with self-healable ability. *Gels (Basel)* 9: 772.

16 Hajidariyor, T., Nuntawad, N., Somsaen, P. et al. (2023). Cryo-induced cellulose-based nanogel from for antibiotic delivery platform. *Int. J. Mol. Sci.* 24: 1230.

17 Sun, M., Wong, J.Y., Nugraha, B. et al. (2019). Cleavable cellulosic sponge for functional hepatic cell culture and retrieval. *Biomaterials* 201: 16–32.

18 Li, Y.F., Khuu, N., Gevorkian, A. et al. (2017). Supramolecular nanofibrillar thermoreversible hydrogel for growth and release of cancer spheroids. *Angew. Chem. Int. Edit.* 56: 6083–6087.

19 Curvello, R., Kerr, G., Micati, D.J. et al. (2021). Engineered plant-based nanocellulose hydrogel for small intestinal organoid growth. *Adv. Sci.* 8: 2002135.
20 Torres-Rendon, J.G., Femmer, T., De Laporte, L. et al. (2015). Bioactive gyroid scaffolds formed by sacrificial templating of nanocellulose and nanochitin hydrogels as instructive platforms for biomimetic tissue engineering. *Adv. Mater.* 27: 2989–2995.
21 Kong, W.Q., Wang, C.W., Jia, C. et al. (2018). Muscle-inspired highly anisotropic, strong, ion-conductive hydrogels. *Adv. Mater.* 30: 1801934.
22 De France, K.J., Yager, K.G., Chan, K.J.W. et al. (2017). Injectable anisotropic nanocomposite hydrogels direct in situ growth and alignment of myotubes. *Nano Lett.* 17: 6487–6495.
23 He, X.L., Yang, Y.Q., Fan, J.Y. et al. (2023). Self-repairable ionic skin based on multiple dynamic bonds with self-adhesive, anti-freezing, and antimicrobial capabilities for monitoring human motions. *Adv. Mater. Technol.* 8: 2300710.
24 Cai, C., Zhu, H.M., Chen, Y.J. et al. (2023). Platelet-rich plasma composite organohydrogel with water-locking and anti-freezing to accelerate wound healing. *Adv. Healthc. Mater.* 12: 2301477.
25 Ganguly, K., Dutta, S.D., Randhawa, A. et al. (2023). Transcriptomic changes toward osteogenic differentiation of mesenchymal stem cells on 3D-printed GelMA/CNC hydrogel under pulsatile pressure environment. *Adv. Healthc. Mater.* 12: 2202163.
26 Guan, Q.F., Han, Z.M., Zhu, Y.B. et al. (2021). Bio-inspired lotus-fiber-like spiral hydrogel bacterial cellulose fibers. *Nano Lett.* 21: 952–958.
27 Schmocker, A., Khoushabi, A., Frauchiger, D.A. et al. (2016). A photopolymerized composite hydrogel and surgical implanting tool for a nucleus pulposus replacement. *Biomaterials* 88: 110–119.
28 Chayanun, S., Soufivand, A.A., Faber, J. et al. (2023). Reinforcing tissue-engineered cartilage: nanofibrillated cellulose enhances mechanical properties of alginate dialdehyde-gelatin hydrogel. *Adv. Eng. Mater.* 26: 2300641.
29 Zhong, L., Zhang, Y.H., Liu, F. et al. (2023). Muscle-inspired anisotropic carboxymethyl cellulose-based double-network conductive hydrogels for flexible strain sensors. *Int. J. Biol. Macromol.* 248: 125973.
30 Sun, W.Q., Liu, X.Y., Hua, W.H. et al. (2023). Self-strengthening and conductive cellulose composite hydrogel for high sensitivity strain sensor and flexible triboelectric nanogenerator. *Int. J. Biol. Macromol.* 248: 125900.
31 Han, X.W., Wang, Z.X., Zhou, Z.J. et al. (2023). Aldehyde modified cellulose-based dual stimuli responsive multiple cross-linked network ionic hydrogel toward ionic skin and aquatic environment communication sensors. *Int. J. Biol. Macromol.* 252: 126533.
32 Liu, X.L., Shi, H.Y., Song, F.F. et al. (2024). A highly sensitive and anti-freezing conductive strain sensor based on polypyrrole/cellulose nanofiber crosslinked polyvinyl alcohol hydrogel for human motion detection. *Int. J. Biol. Macromol.* 257: 128800.
33 Danmatam, N., Pearce, J.T.H., and Pattavarakorn, D. (2023). Intelligent self-healable electroactive carboxymethyl cellulose hydrogel containing

conductive polythiophene and acid hydrolyzed cellulose. *J. Appl. Polym. Sci.* 141: e54751.

34 Shi, H.D., Deng, Y.X., and Shi, Y. (2023). Cellulose-based stimuli-responsive anisotropic hydrogel for sensor applications. *ACS Appl. Nano Mater.* 6: 11524–11530.

35 Huang, C.B., Soenen, S.J., van Gulck, E. et al. (2012). Electrospun cellulose acetate phthalate fibers for semen induced anti-HIV vaginal drug delivery. *Biomaterials* 33: 962–969.

36 Allafchian, A., Hosseini, H., and Ghoreishi, S.M. (2020). Electrospinning of PVA-carboxymethyl cellulose nanofibers for flufenamic acid drug delivery. *Int. J. Biol. Macromol.* 163: 1780–1786.

37 Liu, Y., Wang, Q., Lu, Y. et al. (2020). Synergistic enhancement of cytotoxicity against cancer cells by incorporation of rectorite into the paclitaxel immobilized cellulose acetate nanofibers. *Int. J. Biol. Macromol.* 152: 672–680.

38 Lago, B., Brito, M., Almeida, C.M.M. et al. (2023). Functionalisation of electrospun cellulose acetate membranes with PEDOT and PPy for electronic controlled drug release. *Nanomaterials (Basel)* 13: 1493.

39 Amani, M., Rakhshani, A., Maghsoudian, S. et al. (2023). pH-sensitive bilayer electrospun nanofibers based on ethyl cellulose and Eudragit S-100 as a dual delivery system for treatment of the burn wounds: preparation, characterizations, and in-vitro/in-vivo assessment. *Int. J. Biol. Macromol.* 249: 126705.

40 Chen, Y.J., Qiu, Y.Y., Chen, W.B.F., and Wei, Q.F. (2020). Electrospun thymol-loaded porous cellulose acetate fibers with potential biomedical applications. *Mater. Sci. Eng. C* 109: 110536.

41 Wang, T.T., Ke, H.Z., Chen, S.P. et al. (2021). Porous protoporphyrin IX-embedded cellulose diacetate electrospun microfibers in antimicrobial photodynamic inactivation. *Mater. Sci. Eng. C* 118: 111502.

42 Liu, F.G., Li, X.Z., Wang, L. et al. (2020). Sesamol incorporated cellulose acetate-zein composite nanofiber membrane: an efficient strategy to accelerate diabetic wound healing. *Int. J. Biol. Macromol.* 149: 627–638.

43 Jatoi, A.W., Ogasawara, H., Kim, I.S., and Ni, Q.Q. (2020). Cellulose acetate/multi-wall carbon nanotube/Ag nanofiber composite for antibacterial applications. *Mater. Sci. Eng. C* 110: 110679.

44 Jatoi, A.W., Kim, I.S., Ogasawara, H., and Ni, Q.Q. (2019). Characterizations and application of CA/ZnO/AgNP composite nanofibers for sustained antibacterial properties. *Mater. Sci. Eng. C* 105: 110077.

45 Yang, J.K., Wang, K., Yu, D.G. et al. (2020). Electrospun Janus nanofibers loaded with a drug and inorganic nanoparticles as an effective antibacterial wound dressing. *Mater. Sci. Eng. C* 111: 110805.

46 Ramanathan, G., Sobhanadhas, L.S.S., Jeyakumar, G.F.S. et al. (2020). Fabrication of biohybrid cellulose acetate-collagen bilayer matrices as nanofibrous spongy dressing material for wound-healing application. *Biomacromolecules* 21: 2512–2524.

47 Nada, A.A., Ali, E.A., Soliman, A.A.F. et al. (2020). Multi-layer dressing made of laminated electrospun nanowebs and cellulose-based adhesive for comprehensive wound care. *Int. J. Biol. Macromol.* 162: 629–644.

48 Ruhela, A., Kasinathan, G.N., Rath, S.N. et al. (2021). Electrospun freestanding hydrophobic fabric as a potential polymer semi-permeable membrane for islet encapsulation. *Mater. Sci. Eng. C* 118: 111409.

49 Abdulmalik, S., Gallo, J., Nip, J. et al. (2023). Nanofiber matrix formulations for the delivery of Exendin-4 for tendon regeneration: *in vitro* and *in vivo* assessment. *Bioact. Mater.* 25: 42–60.

50 Luo, H.L., Gan, D.Q., Gama, M. et al. (2020). Interpenetrated nano- and submicro-fibrous biomimetic scaffolds towards enhanced mechanical and biological performances. *Mater. Sci. Eng. C* 108: 110416.

51 Lee, J., Moon, J.Y., Lee, J.C. et al. (2021). Simple conversion of 3D electrospun nanofibrous cellulose acetate into a mechanically robust nanocomposite cellulose/calcium scaffold. *Carbohydr. Polym.* 253: 117191.

52 Patel, D.K., Dutta, S.D., Hexiu, J. et al. (2020). Bioactive electrospun nanocomposite scaffolds of poly(lactic acid)/cellulose nanocrystals for bone tissue engineering. *Int. J. Biol. Macromol.* 162: 1429–1441.

53 Mohammadalipour, M., Behzad, T., Karbasi, S. et al. (2023). Osteogenic potential of PHB-lignin/cellulose nanofiber electrospun scaffold as a novel bone regeneration construct. *Int. J. Biol. Macromol.* 250: 126076.

54 Gonçalves, A.M., Leal, F., Moreira, A. et al. (2023). Potential of electrospun fibrous scaffolds for intestinal, skin, and lung epithelial tissue modeling. *Adv. Nanobiomed. Res.* 3: 2200104.

55 Szewczyk, P.K., Berniak, K., Knapczyk-Korczak, J. et al. (2023). Mimicking natural electrical environment with cellulose acetate scaffolds enhances collagen formation of osteoblasts. *Nanoscale* 15: 6890–6900.

56 Wang, J.P., Chiappone, A., Roppolo, I. et al. (2018). All-in-one cellulose nanocrystals for 3D printing of nanocomposite hydrogels. *Angew. Chem. Int. Edit.* 57: 2353–2356.

57 Guo, Z.Q., Ma, C.D., Xie, W.G. et al. (2023). An effective DLP 3D printing strategy of high strength and toughness cellulose hydrogel towards strain sensing. *Carbohydr. Polym.* 315: 121006.

58 Chen, Y., Wang, L.Y., Wang, Y., and Zhou, Y.G. (2023). Microtube embedded hydrogel bioprinting for vascularization of tissue-engineered scaffolds. *Biotechnol. Bioeng.* 120: 3592–3601.

59 Schaffner, M., Rühs, P.A., Coulter, F. et al. (2017). 3D printing of bacteria into functional complex materials. *Sci. Adv.* 3: aao6804.

60 Shin, S., Kwak, H., Shin, D., and Hyun, J. (2019). Solid matrix-assisted printing for three-dimensional structuring of a viscoelastic medium surface. *Nat. Commun.* 10: 4650.

61 Fan, Y.C., Yue, Z.L., Lucarelli, E., and Wallace, G.G. (2020). Hybrid printing using cellulose nanocrystals reinforced GelMA/HAMA hydrogels for improved structural integration. *Adv. Healthc. Mater.* 9: 2001410.

62 Cheng, F., Cao, X., Li, H.B. et al. (2019). Generation of cost-effective paper-based tissue models through matrix-assisted sacrificial 3D printing. *Nano Lett.* 19: 3603–3611.

63 Rad, R.M., Daul, B., Glass, P. et al. (2023). 3D printed magnet-infused origami platform for 3D cell culture assessments. *Adv. Mater. Technol.* 8: 2202204.

64 Slejko, E.A., Gorella, N.S., Gasparini, R. et al. (2023). Tailoring 3D printed cellulose acetate properties produced via direct ink writing: densification through over-extrusion and evaporation rate control. *Polym. Eng. Sci.* 63: 3786–3797.

65 Boons, R., Siqueira, G., Grieder, F. et al. (2023). 3D bioprinting of diatom-laden living materials for water quality assessment. *Small* 19: 2300771.

66 Nagel, Y., Sivaraman, D., Neels, A. et al. (2023). Anisotropic, strong, and thermally insulating 3D-printed nanocellulose-PNIPAAM aerogels. *Small Struct.* 4: 2300073.

3

Sources, Structures, and Properties of Hyaluronic Acid

Hyaluronic acid, also known as hyaluronan, is a kind of anionic polysaccharide with helical structure, which is widely found in connective and epithelial tissues (more than half of hyaluronic acid in human skin) [1]. It plays an important role in maintaining tissue lubrication and elasticity [2]. At present, some studies have found that microbial fermentation can also produce hyaluronic acid [3].

Hyaluronic acid is composed of D-glucuronic acid and *N*-acetylaminoglucose alternately linked by β-(1-4) and β-(1-3) glucosidic bonds. The hydrophilic property of hyaluronic acid mainly comes from its large amount of carboxyl and hydroxyl groups, and the hydrophobic domain composed of CH groups in the axial direction of hyaluronic acid makes hyaluronic acid amphiphilic. In the 1980s, Pharmacia applied hyaluronic acid to the medical field for the first time, which opened up a new research direction in the biomedical field. The molecular weight of hyaluronic acid directly affects its mechanical and biological properties [4]. For example, high-molecular-weight hyaluronic acid ($>1 \times 10^6$ kDa) has anti-inflammatory and anti-angiogenic effects, while low- and medium-molecular-weight (4×10^5 kDa) hyaluronic acid can stimulate inflammatory cells to release cytokines [5]. The relationship between hyaluronic acid and cells depends not only on its molecular weight, but also on the related cellular receptors. A large number of studies have shown that hyaluronic acid can bind to CD44 receptors to regulate cell behavior, based on which hyaluronic acid is also used in targeted therapy of tumor cells [6, 7]. At a neutral pH value, the hydroxyl and acetyl amine groups contained in hyaluronic acid can form a stable secondary structure with water molecules, and hyaluronic acid will be hydrolyzed when $pH < 4$ or $pH > 11$ [8, 9].

At present, the design of hyaluronic acid derivatives is mainly divided into two categories: one is “monolithic,” which modifies the terminal group of hyaluronic acid to make it perform its biological function as a whole; the other is “living,” which enables hyaluronic acid to form covalent bonds with cells or tissues in different microenvironments.

Natural Polymers for Biomedical Applications, First Edition. Wenguo Cui and Lei Xiang.

3.1 Hyaluronic-Acid-Based Hydrogel for Biomedical Application

3.1.1 Cell and Organoids Culture

During tissue development, the microenvironment gradually changes from an environment rich in cell–cell interactions to an environment dominated by cell–ECM (extracellular matrix) interactions. In order to further study the effect of this microenvironmental change on cell behavior, Cosgrove et al. [10] developed a 2D hydrogel to make the N-cadherin "HAVDI" adhesion sequence (that mimics the cell-to-cell interaction). (Interaction) and the "RGD" adhesion sequence of fibronectin (imitating the interaction between cells and extracellular matrix) can be co-presented independently, while independently regulating matrix stiffness. The decoupling of these signals shows that HAVDI (in the context of constant RGD) reduces the contraction state of MSCs and the localization of nuclear YAP/TAZ, resulting in a change in the interpretation of ECM stiffness, thereby changing the proliferation and differentiation of downstream cells. ECM provides cells with a complex and dynamic microenvironment, characterized by various physical and biochemical signals. The extent to which these ECM signals regulate cell behavior depends largely on dimensionality. For example, there are significant differences in the expression of chondrogenic genes in articular chondrocytes in 2D and 3D culture, and the influence of dimensionality on the phenotype of other cell types has also been observed. How to better study the effects of the three-dimensional environment on various cells in the cell culture system has become a hot research topic (Figure 3.1).

Vega et al. [11] used a photo-mediated reaction between a norbornene-functionalized HA macromolecule and a dithiol cross-linker to form a hydrogel and encapsulate cells. Then, the unreacted norbornene in the hydrogel reacts with the monothiolated polypeptide through light exposure controlled by a sliding opaque mask. This method is used to introduce single or multiple gradient peptides to better simulate the interaction between cells and between cells and matrix. In this environment, cells are easy to image to obtain quantifiable outcome.

Similarly, Tam et al. [12] developed photosensitive hyaluronic acid hydrogels, which can be activated by single-photon or two-photon radiation to immobilize a specific volume of biomolecules in 3D hydrogels. By controlling the spatial position of protein immobilization, a cell culture environment with different protein concentration gradients can be created in these hydrogels. Based on the nature of breast cancer cells that can secrete metalloproteinases, they chemically cross-linked hyaluronic acid with polypeptides that can be degraded by metalloproteinases. The hydrogel prepared by this method can allow breast cancer cells to degrade hydrogels according to their own invasion needs. By independently adjusting the chemical and physical properties of these hydrogels, various factors that affect tumor cell invasion can be further explored. Cell–matrix adhesion interaction can regulate stem cell differentiation, and its underlying mechanism, especially the mechanism of direct 3D encapsulation in the hydrogel. Khetan et al. [13]

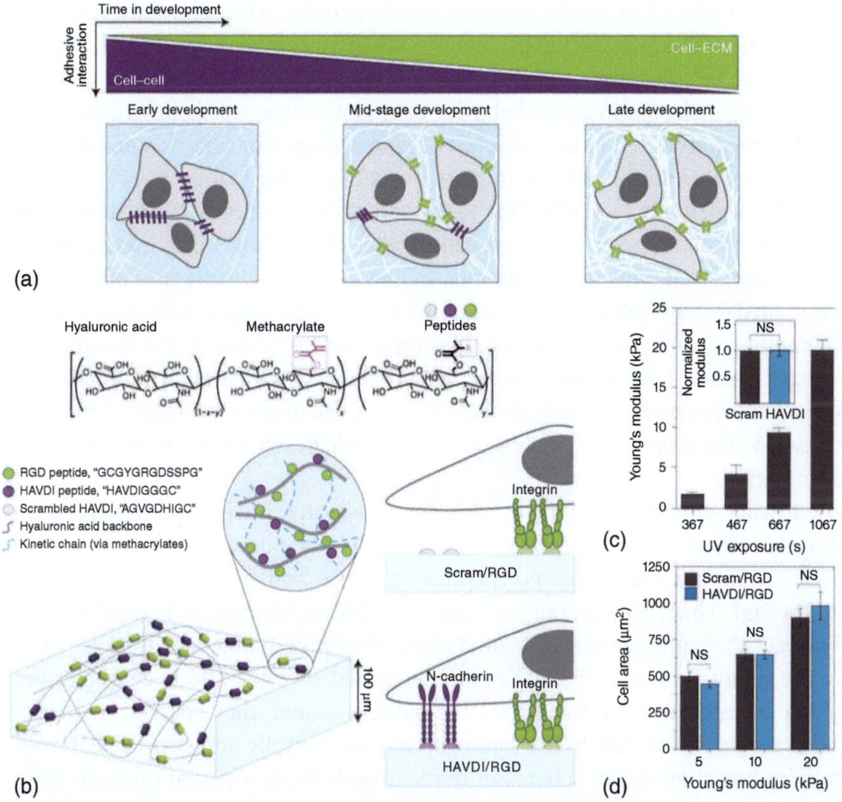

Figure 3.1 (a) Schematic diagram depicting the evolution of the mechanical properties of the extracellular matrix during mesenchymal development. (b) The preparation of hydrogels using UV photocross-linking of hyaluronate and adhesion proteins is illustrated. (c) The mechanical properties of hydrogels were modulated by varying the duration of UV irradiation and the elastic modulus of the gels was characterized by AFM. Different matrix components allowed different types of cell–matrix interactions. (d) The effect of different rigid gels on the morphology of mesenchymal stem cells was evaluated. Source: Ref. [10]/Springer Nature.

further studied the influence of the internal traction of the covalently cross-linked hyaluronic acid hydrogel that can be degraded by cells on cell differentiation. They found that allowing HMSCs-mediated degradation of the hydrogel exhibited a high degree of cell spreading and high traction, and facilitated the differentiation of cells to osteoblasts. In addition, switching the allowable hydrogel to a restricted state by delaying the secondary cross-linking can further reduce the degradation of the hydrogel, inhibit traction, and cause the transition from osteogenesis to lipid formation without changing the expanded cell morphology.

3.1.2 Cell Behaviors Regulation

Cell migration plays an important role in many physiological and biological processes. It is not only affected by the physical and chemical properties of the surrounding matrix, but also by signal gradients generated by adjacent/distal cells. Yu et al. [14] used matrix metalloproteinases sensitive peptides to cross-link HAMA macromolecules to prepare a simulated dynamic ECM. Under the mediation of inflammatory cells, the invasion behavior of smooth muscle cells in hydrogel was studied. They found that in the presence of macrophage-like cells as a non-contact inducer, smooth muscle cell can migrate deeper into the hydrogel (120 μm) and have higher sensitivity to MMPs. Due to the partial degradation of the hydrogel and chemokine gradient, smooth muscle cell invades the hydrogel in a mesenchymal manner. Compared with the continuous induction of U937 cells, smooth muscle cell migrated deeper (180 μm) in the degradable MMPs hydrogel, indicating that the degradability of the hydrogel has a greater impact on cell invasion. The results of in vivo experiments showed that cells can invade the hydrogel, and as the content of MMP-sensitive peptides in the hydrogel increases, the invasion distance of the cells increases. Various signaling proteins play a huge role in cell adhesion, proliferation, and migration.

Wieduwild and Howard [15] provide new ideas for integrating proteins into hydrogel scaffolds more accurately and efficiently. They used a remote-controlled SpyTag-containing elastin-like polypeptide to derive and modify hyaluronic acid (HA-SpyTag). Then, HA-SpyTag is mixed with TriCatcher protein to quickly form a hydrogel through spontaneous amidation. This programmed scaffold can test how the interaction of individual matrix-anchored proteins affects cell behavior. In order to verify the applicability of the culture system, they coupled the extracellular region of EpCAM to the hydrogel and successfully adjusted the behavior of human breast epithelial spheres in the hydrogel. Oxygen partial pressure plays an important role in both angiogenesis and angiogenesis, and the angiogenesis process has a great influence on the angiogenesis of endothelial cells. Hsu et al. [16] systematically studied different concentrations of hyaluronic acid added the formation of a three-dimensional endothelial cell network of fibrinogen hydrogel under three uniform oxygen partial pressures (normal oxygen, 5%, and 1% O_2) and two oxygen gradients (normal oxygen and 5% O_2). They found both the oxygen concentration and the spatial distribution of oxygen play a key role in regulating the formation of endothelial cell networks. Interestingly, the presence of oxygen gradient promotes the proliferation and networking of endothelial cells in the hydrogel. In addition, the formed endothelial cell network tends to align along the gradient direction. In terms of matrix composition: although the addition of hyaluronic acid has little effect on cell proliferation, it can stimulate endothelial cells to form a network in the hydrogel. Combined with the addition of oxygen gradient and hyaluronic acid, the formation of 3D endothelial cell network can be further enhanced, which indicates their synergy and also implies the key role of oxygen gradient. Many 3D in vitro models induce the formation of breast cancer spheroids. However, this cannot generalize the complex phenotype alone in the body. In order to effectively screen therapeutic drugs, it is urgent to compare the gold standard of xenotransplantation to validate the in vitro tumor spheroid model. Baker et al. [17] used fast-reacting HA-aldehyde and slow-reacting HA ketones in combination with PEG–hydroxylamine to prepare a novel oxime cross-linked

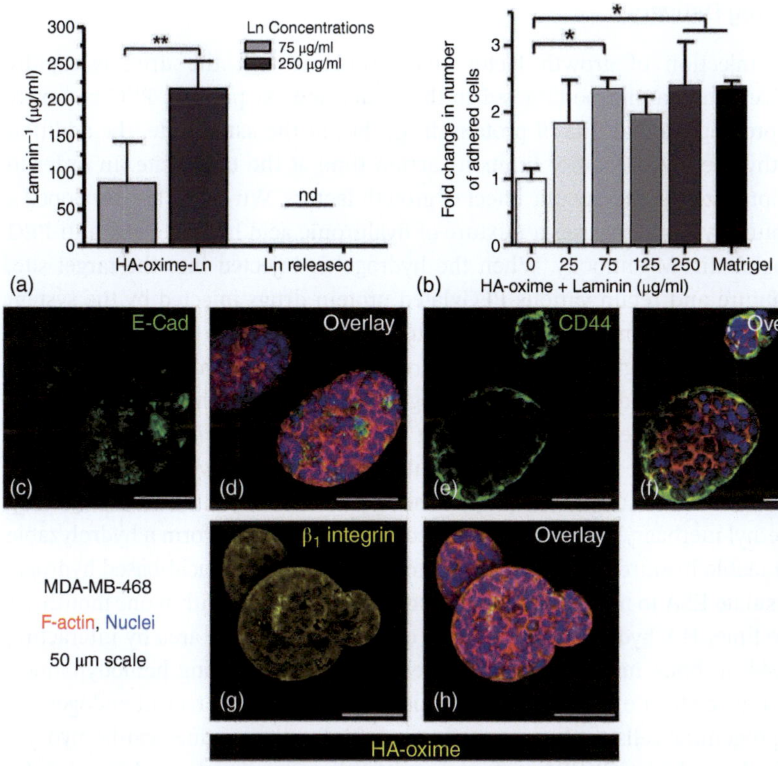

Figure 3.2 (a) The amount of laminin retained in the hyaluronic acid oxidized gel was assessed. (b) The effect of laminin-containing hyaluronic acid oxidized gel on the proliferation of breast cancer cells was compared. (c–h) Immunocytochemical staining to observe the expression of breast cancer cells in encapsulated hyaluronic acid oxidized gel. Source: Ref. [17]/John Wiley & Sons.

hyaluronic acid hydrogel. The oxime linkage is stable during the hydrolysis process, so breast cancer cells can be embedded for a long time. This is a comparison with strategies that are inherently limited by reversible reactions such as using hydrazone or Diels–Alder chemistry for cross-linking. At the same time, oxime chemistry is not sensitive to oxidation, is easy to use, and can control the gelation rate. And compared with the current gold standard for transplanted tumors in mice, benchmark the gene expression of breast cancer cells grown in these hyaluronic acid–oxime hydrogels, and compare them with conventional culture in 3D Matrigel and 2D TCP to evaluate drugs reaction. This hyaluronic acid–oxime breast cancer model is based on the pan-oncogene expression profiles (including 730 genes) of three different human breast cancer subtypes and maintains the gene expression profiles most similar to xenogeneic tumors. Through gene set mutation analysis, the difference in gene expression of breast cancer cultures in the 12 typical pathways of hyaluronic acid–oxime and Matrigel or 2D was confirmed. Importantly, the drug response depends on the culture method. Compared with Matrigel, breast cancer cells respond better to RAC inhibitors (EHT-1864) and PI3K inhibitors (AZD6482) (Figure 3.2).

3.1.3 Drug Delivery

Systemic injection of growth factors has problems that the drug is rapidly metabolized and cannot be targeted to the injury site. At present, PEG is mostly used to prolong the half-life of protein drugs, but at the same time, the addition of polyethylene glycol further limits its action time at the target site. In order to further optimize the therapeutic effect of growth factors, Wu et al. [18] developed a drug capture system that uses a mixture of hyaluronic acid hydrogel and anti-PEG immunoglobulin M antibody. When the hydrogel is injected into the target site, it can capture and retain various PEGylated protein drugs injected by the system at that site, so as to make up for the insufficient time of the PEGylated protein drugs in the target region. Also, for the drug delivery problem of growth factors, Shamskhou et al. [19] developed the hydrogel system based on hyaluronic acid and heparin delivers interleukin 10 by utilizing the ability of heparin to reversibly bind interleukin 10, which further improves interleukin 10 effective in reducing pulmonary fibrosis. MacArthur et al. [20] chemically modified hyaluronic acid by using hydroxyethyl methacrylate, and cross-linked by free radicals to form a hydrolyzable and degradable hyaluronic acid hydrogel, and used hyaluronic-acid-based hydrogel to encapsulate ESA to maintain the release of the polypeptide within one month. At the same time, HA hydrogel can limit the expansion of infarct area by interacting with CD44 on bone marrow progenitor cells, thereby improving hemodynamics. The synergistic effect of ESA and hyaluronic acid hydrogel can recruit endogenous cardiac progenitor cells to the marginal zone of infarction, reduce cardiomyocyte apoptosis through the CXCR4 pathway, and further improve hemodynamics. By mixing β-cyclodextrin (CD, host)- or adamantane (AD, guest)-modified hyaluronic acid, Wang et al. [21] prepared a guest–host HA with injectable and self-healing ability hydrogel system. This system can experiment with local sustained release of miR-302 to promote the proliferation and regeneration of cardiomyocytes after myocardial infarction. The up-regulation of MiR-223 microRNAs (MiRNAs) indicates that the polarization of macrophages tends to be an anti-inflammatory (M2) phenotype. Loading them in hyaluronic acid hydrogel can significantly improve the stability of aerobiological organisms and better induce macrophages cells differentiate to the M2 phenotype, thereby reducing inflammation and promoting epidermal regeneration [22].

The controlled release system of macromolecular drugs based on various hydrogels (such as various protein drugs) has been extensively studied, but the delivery of small-molecule drugs and bioactive ions is still a challenge. Zhang et al. [23] developed bioactive nanocomposite hydrogels of hyaluronic acid and self-assembled magnesium pamidronate nanoparticles for local elution and simultaneous release of bioactive ions and small-molecule drugs on demand. Through the dynamic coordination of Mg^{2+} and PAM, the hydrogel has excellent injectability so that the hydrogel can adapt to irregular-shaped bone defects. The magnesium ions released from the hydrogel promote the osteogenic differentiation of encapsulated HMSCs and the activation of alkaline phosphatase. The self-assembled nanostructure of PAM-Mg^{2+} in the hydrogel improves the stability of the dexamethasone prodrug in the hydrogel

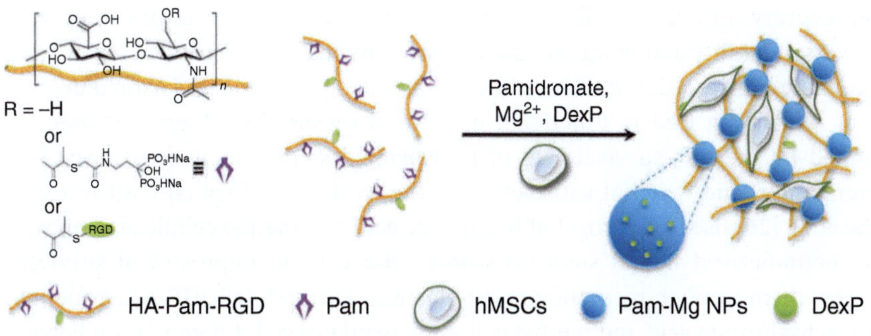

Figure 3.3 Schematic illustrations of the fabrication of self-assembled HA-Pam-Mg nanocomposite hydrogels. Source: Ref. [23]/John Wiley & Sons/CC BY 4.0.

network through the interaction of phosphate and Mg^{2+}, and the activated alkaline phosphatase then activates dexamethasone the dephosphorylation of the prodrug of methasone and accelerate the release of dexamethasone from the hydrogel for further promoting the osteogenesis of HMSCs. This positive feedback circuit that controls the activation and release of dexamethasone significantly enhances bone regeneration at the hydrogel implant site (Figure 3.3).

3.1.4 Tissue Regeneration

Thanks to the unique physical and chemical properties and biological activities of hyaluronic acid, hydrogels based on hyaluronic acid as the main active ingredient are also widely used in nerve repair, wound healing, and vascular regeneration.

Blood vessels are an important channel for tissues to provide nutrients and remove metabolic wastes. It directly affects the results of tissue repair and regeneration engineering. Precisely controlling the formation of local microvessels is of great significance to tissue engineering. Inspired by the dry curl of apple peel, Zhang et al. [24] used photocross-linked HAMA to reduce the shrinkage between the upper and lower layers. Driven by the difference in expansion rate, the microvascular structure was prepared. Microvascular structures with a diameter of 50–500 μm can be prepared by adjusting the concentration of the hydrogel and the photocross-linking time. In animal experiments, the prepared blood vessel structure significantly reduced the necrotic area of the flap and increased the blood vessel density of the flap. In addition, Kang et al. [25] used the pressure field formed by the residual surface acoustic wave to manipulate the properties of particles with high resolution in a non-invasive manner, and developed a new blood vessel formation method. They used hyaluronic acid hydrogel as a matrix and added catecholamines to further promote angiogenesis and the secretion of anti-inflammatory paracrine factors. They injected the hyaluronic acid–catecholamines solution mixed with the cells into the chamber, and by applying surface acoustic waves, the cells in the solution were molded into side branch cylinders at certain intervals in the horizontal and vertical directions (X and Z directions). SSAW-induced cell aggregation enhances the secretion of VEGF and IL-10, thereby significantly enhancing angiogenesis and tissue recovery. In addition, the patterned HUVEC/hADSC co-culture improved blood vessel stability and maturity, and produced perfused microvessels.

Traumatic injury to the spinal cord can lead to demyelination and axonal degeneration, which can lead to functional motor and sensory loss. Stem cell therapy is limited by the low survival rate of transplanted stem cells and uncontrolled differentiation, and it is not satisfactory in the treatment of spinal cord trauma. Mothe et al. [26] used a hydrogel of hyaluronic acid and methyl cellulose to encapsulate brain-derived neural stem cells/progenitor cells to improve cell survival, and grow them through recombinant rat platelets. Factor-A (rPDGF-A) covalently modifies hyaluronic acid and methyl cellulose, regulates cell differentiation behavior, and promotes its differentiation into oligodendrocytes. This stem cell transport strategy significantly improves the survival rate of the graft, promotes the targeted differentiation of the cells, and retains the oligodendrocytes and neurons around the lesion to the maximum. In addition to the physical injury of traumatic spinal cord injury, some patients will also experience severe inflammation in the subarachnoid space or subarachnoid space, leading to the formation of scars and dead spaces at the injury site. This situation is called trauma posterior syringomyelia. High-molecular-weight hyaluronic acid plays a role in inflammation and tissue repair by interacting with inflammatory cells and ECM proteins. When the physical mixture of hyaluronic acid and methylcellulose is injected into the intrathecal space (the fluid-filled surrounding spinal cord), it can alleviate the inflammatory response after spinal cord injury and degrade/dissolve after four to seven days. Hyaluronic acid and methyl cellulose also reduced IL-1a cytokine expression, and promoted the recovery of synaptic conduction function in rats. The pain caused by nerve damage seriously affects the patient's quality of life, and the changes in intracellular Ca^{2+} levels are thought to affect cell signaling and downstream processes. Tay et al. [27] further studied the control of calcium influx channels through non-invasive neuromodulation. They used 4-arm PEG vinyl sulfone with high-molecular-weight (700 kDa) hyaluronic acid–thiol and 1 μm diameter fluorescent magnetic particles MMPs to react with thiol to synthesize a magnetic hyaluronic acid gel. Membrane-bound magnetic nanoparticles can stretch the lipid bilayer to increase the opening probability of endogenous mechanically sensitive N-type Ca^{2+} channels, thereby inducing calcium influx in the cortical neural network. They were surprised to find that acute magneto-mechanical stimulation mainly induces calcium influx in DRG neurons through the endogenous mechanically sensitive TRPV4 and Piezo2 channels. Next, using the receptor adaptation properties of primary rat dorsal root neurons for chronic magneto-mechanical stimulation, it was found that it reduced the expression of Piezo2 channels, further confirming the reliability of reducing pain transmission in patients by adjusting mechanically sensitive channels (Figure 3.4).

High intraocular pressure caused by glaucoma is the primary cause of irreversible blindness. At present, there are two main treatment measures for high intraocular pressure: one is to reduce the production of aqueous humor and promote the outflow of aqueous humor through the use of eye drops. The second is to expand the outlet of aqueous humor through surgery. Aqueous humor can not only be discharged

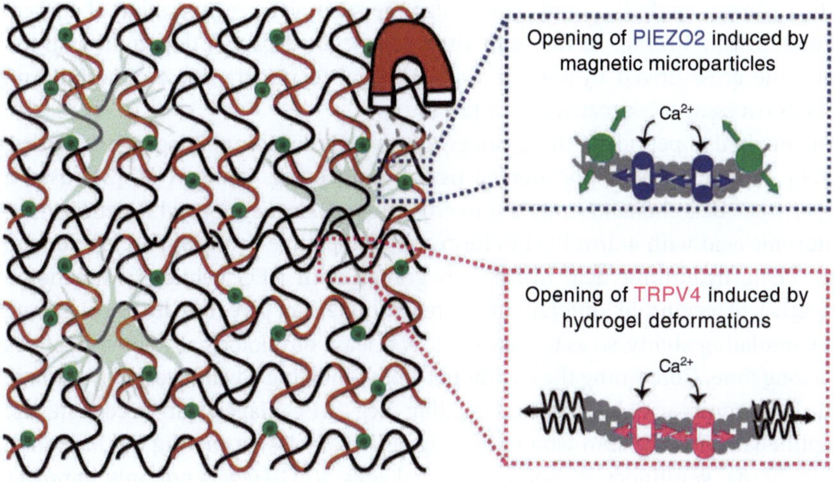

Figure 3.4 Magnetic particles embedded in hydrogels can activate PIEZO2 mechanosensitive channels by membrane stretching. On the other hand, magnetic-induced deformation of the gel could activate TRPV4 mechanosensitive channels. Source: Ref. [27]/John Wiley & Sons.

from the conventional way of trabecular meshwork, but it can also be discharged through the extracellular matrix of the ciliary muscle into the suprachoroidal space. Hyaluronic acid is a type of polysaccharide substance that exists widely in the eye. Based on this, Chae et al. [28] developed an injectable hyaluronic acid hydrogel to expand the upper choroidal space. The goal of treatment to reduce intraocular pressure for four months. Clinically, severe eye diseases or cataracts caused by diabetes require lens removal or replacement surgery. At the same time, it is necessary to supplement the vitreous humor to maintain eyeball volume, support and protect the retina. At present, silicone oil and fluorinated gas are commonly used as intraocular fillers, but these fillers have long-term cytotoxicity and emulsification tendency. Hyaluronic acid is biologically similar to natural vitreous humor, but its relatively fast degradation rate limits its application as a substitute for vitreous humor. Raia et al. [29] used horseradish peroxidase and hydrogen peroxide to combine hyaluronic acid and silk fibroin through enzymatic cross-linking, and controlled the degradation speed of hyaluronic acid while retaining good biocompatibility. By adjusting the concentration of hydrogen peroxide and the content of silk fibroin, the mechanical properties and swelling properties of the hydrogel can be further adjusted, revealing the great potential of this type of hydrogel as a vitreous humor substitute.

Corneal damage will affect the corneal epithelium and corneal stroma at the same time. Realizing the controllable regeneration of corneal epithelium and stroma is the key to reducing blindness caused by corneal damage. In the past, corneal implants often needed to be sutured due to lack of adhesion, but this action would increase the chance of inflammation and cause treatment failure. Koivusalo et al. [30] wrapped adipose-derived stem cells in a hydrogel, then combined with thiolated type IV collagen or laminin peptides, and planted a limbus on the surface of the hydrogel epithelial stem cells realize the synergistic regeneration of corneal epithelium and stroma. At the same time, they grafted part of dopamine into the hydrazone cross-linked hyaluronic acid hydrogel to impart tissue adhesion and avoid unnecessary risks caused by sutures.

The method of peptide grafting can give hydrogels different biological properties, thereby expanding its application in tissue engineering. Wang et al. [31] used a copper-free click chemistry method to cross-link dibenzocyclooctyl functionalized hyaluronic acid with 4-arm PEG to form a biocompatible hyaluronic acid hydrogel glue. Then, hyaluronic acid hydrogel is grafted with REG peptide (a functional derivative of erythroid differentiation regulator 1) to give the hydrogel strong cell-stimulating ability, so as to continuously release physiologically active peptides for a long time. Combining the traditional wound healing advantages of hyaluronic acid, hyaluronic acid hydrogel embedding Reg (REG-Hagel) can accelerate the re-epithelialization in skin wound healing, especially by promoting the migration of fibroblasts, keratinocytes, and endothelial cells. REG-Hagels not only improves the speed of wound healing, but also improves the quality of wound healing, which is embodied in the preservation of good biocompatibility, which induces higher collagen deposition and more capillary formation. Hydrogen sulfide, as a gas messenger, has the potential to induce macrophages to differentiate into the M2 phenotype. Wu et al. [32] used the biomimetic hyaluronic acid hydrogel formed in situ as the matrix and mixed with the pH-controllable hydrogen sulfide donor JK1 to form a new type of hyaluronic acid-JK1 hybrid system. Hyaluronic acid-JK1 hydrogel is an ideal JK1 release scaffold, and its hydrogen sulfide release curve is prolonged with changes in pH. In vitro studies have shown that JK1 can induce phenotypic polarization of macrophages M2. Animal experiments have shown that hyaluronic acid–JK1 hybrid hydrogel can significantly promote wound re-epithelialization, collagen deposition, angiogenesis, and cell proliferation.

3.2 Hyaluronic-Acid-Based Electrospinning for Biomedical Application

3.2.1 Drug Delivery and Antibacterial

Hyaluronic acid is a highly hydrophilic natural polysaccharide, which can effectively remove various exudates produced during wound healing to keep the wound dry and create favorable external conditions for wound healing. In addition to their good antibacterial ability, silver ions can also act as an antioxidant to remove large amounts of reactive oxygen species produced when macrophages and neutrophils invade the wound. Based on this, El-Aassar et al. [33] used polygalacturonic acid to reduce silver ions into nanoparticles while increasing the stability of silver nanoparticles, which will be embedded in polygalacturonic acid and hyaluronic acid. Silver nanoparticles are prepared into nanofiber mats by electrospinning. This new type of wound adjuvant can effectively inhibit wound inflammation and accelerate the wound healing process. Although hyaluronic acid is highly soluble in water, most of

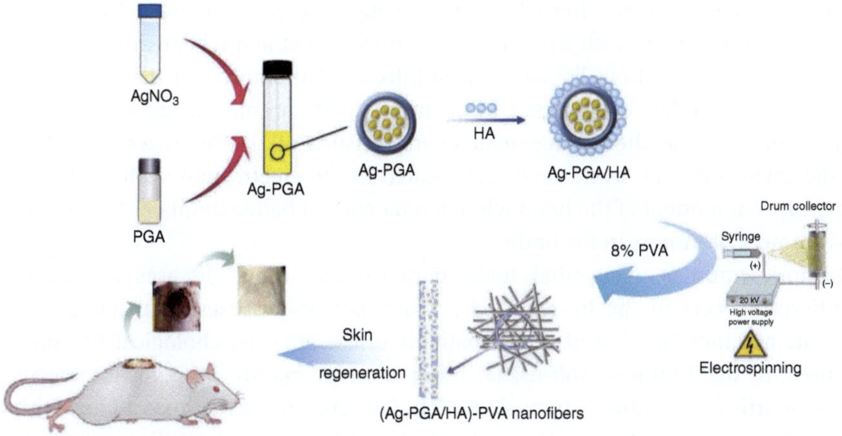

Figure 3.5 Diagram illustrating the formation of (Ag-PGA/HA)-PVA. Source: Ref. [33]/with permission of Elsevier.

them are electrospinning by preparing an organic hyaluronic acid solution. This is because the ionicity of hyaluronic acid leads to long-range electrostatic interactions, and the presence of counter ions leads to a sharp increase in the viscosity of the hyaluronic acid aqueous solution, which affects the stability of the electrospinning process (Figure 3.5).

Séon-Lutz et al. [34] further optimized the formulation of the hyaluronic acid aqueous solution to meet the needs of electrospinning technology in order to avoid the potential toxic side effects of organic solutions. They used the good intermolecular interaction between hydroxypropyl-β-cyclodextrin and polyvinyl alcohol and hyaluronic acid to further stabilize the electrospinning performance of the solution. At the same time, the weight ratio of hyaluronic acid:polyvinyl alcohol reaches 1:1, which is much higher than other literature reports. And using naproxen as a model drug, studies have shown that the molecule can easily penetrate into the stent, whether in an aqueous solution or in supercritical CO_2, and exhibits sustained drug release characteristics for more than 48 hours.

3.2.2 Tissue Regeneration

In addition, hyaluronic-acid-based nanofibers represent a new type of bioactive wound dressing, but hyaluronic acid nanofibers alone can hardly meet the mechanical strength required for tissue engineering. Hussein et al. [35] introduced cellulose nanocrystals into hyaluronic acid nanofibers to improve the mechanical properties of the fibers, and the addition of L-arginine further improved the ability to promote wound healing. By optimizing the fiber structure, the mechanical properties can also be enhanced while retaining the original biologically active functions of hyaluronic acid. Movahedi et al. [36] used coaxial electrospinning technology making polyurethane as the fiber core and hyaluronic acid as the fiber outer shell. While improving the performance of fiber mechanics, it effectively promoted the cell adhesion activity of mouse fibroblasts. Also using polycaprolactone and hyaluronic acid electrospun fibers with a putamen structure, Rao et al. [37] constructed an in vitro migration model of glioblastoma multiforme. Different from the commonly used in vitro models constructed by hydrogels, the oriented electrospun fibers can better simulate the arrangement characteristics of the brain white matter. At the same time, the hyaluronic acid located in the electrospun shell is also an important component of the brain white matter since it better simulates the growth environment of tumors in the brain.

Tumors, traumas, congenital malformations, and other diseases can cause soft-tissue defects in the human body, which not only causes great damage to the patient's mobility, but also puts patients under great psychological pressure. At present, the filling of soft-tissue defect areas is mainly through autologous tissue or artificial prosthesis transplantation. However, these methods all have this drawback. For example, autologous tissue transplantation can still cause tissue defects in other parts, and fat transplantation has the problem of poor angiogenesis; prosthetic transplantation can easily trigger the human immune response and cause a series of complications. In order to solve these problems, Li et al. [38] developed a method that combines electrospinning and hydrogel to prepare soft-tissue filling materials. They successfully prepared biomaterials that match the mechanical properties of natural adipose tissue by covalently combining thiolated hyaluronic acid hydrogel and polycaprolactone electrospun fibers. This biomaterial not only meets the mechanical properties required for soft-tissue transplantation, but also has high porosity and cell permeability, which is conducive to early angiogenesis. In addition, Song et al. [39] further explored the influence of the soft and hard characteristics of the electrospun scaffold on the tissue repair process. They adjusted the hardness of the electrospinning scaffold by changing the degree of cross-linking of methacrylated hyaluronic acid. Studies have shown that the softer methacrylated hyaluronic acid fiber network is easily deformed and compacted by cell traction, while the harder hyaluronic acid fiber network has a significant increase in cell migration behavior within a few weeks of implantation in the meniscus. When the stent is sandwiched between the meniscus tissues and implanted under the skin, compared with the soft methacrylated hyaluronic acid fiber network, the harder methacrylated hyaluronic acid fiber network exhibits a positive effect after four weeks (Figure 3.6).

At the same time, Kim et al. [40] adjusted the adhesion of the whole stent by changing the RGD density of hyaluronic acid. They found the proliferation and local adhesion formation of human bone marrow mesenchymal stem cells depend on the density of RGD, and have nothing to do with the degree of softness of the fibers. The expression of cartilage markers is affected by both fiber mechanics and adhesion, where softer fibers and lower RGD density usually promote cartilage formation.

Besides the good biological activity and biocompatibility of hyaluronic acid, hyaluronic acid oligosaccharides show good performance in vascular endothelialization effect. Based on this, Li et al. [41] prepared a biomimetic nanofiber network based on hyaluronic acid oligosaccharide-modified collagen and its mineralization products by the room-temperature self-assembly method. This composite material exhibits some characteristics of natural bone in terms of composition

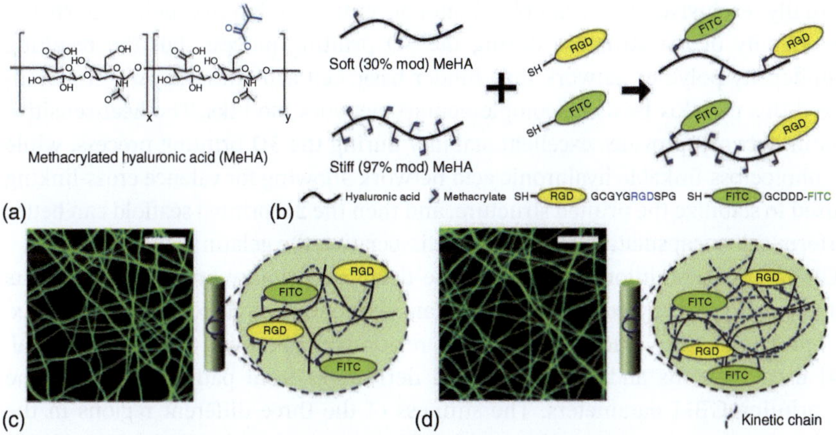

Figure 3.6 (a) Soft and hard methylated hyaluronan were prepared by modifying the hydroxyl groups of hyaluronan to varying degrees with methacrylate. (b) RGDs and fluorescein peptides were attached to methylated hyaluronan. (c) Soft hyaluronan fibrous network consisted of kinetically cross-linked fibers. (d) Hard hyaluronan fibrous network consisted of more kinetically cross-linked fibers. Hard hyaluronan fibrous network was prepared by modification of the hydroxyl groups of hyaluronan with methacrylate. Source: Ref. [39]/John Wiley & Sons.

and microstructure and is applied to the culture of arterial endothelial cells and mouse parietal bone cells. They found that the cells are tightly attached to the nanofibers and penetrate into the material to form an interconnected cell colony. At the same time, the expression of alkaline phosphatase and osteocalcin also increased significantly. The biomimetic tubular structure prepared by electrospinning technology has great potential in inducing blood vessel formation and nerve regeneration. Apsite et al. [42] reported the application of 4D biomanufacturing methods in the manufacture of artificial nerve grafts. They prepared uniaxially arranged polycaprolactone-glyceryl sebacate and randomly arranged HAMA fiber double-layer scaffolds by electrospinning. Based on the composition of each layer with different swelling coefficients in water, this double-layer fiber can spontaneously form a tubular structure. The inner fibers of the formed tubular structure have a specific arrangement direction, which can provide contact guidance for cells and form unidirectional protrusion growth. And by controlling the thickness of the two layers of fibers, the tube diameter can be adjusted to meet the needs of different tissue engineering.

3.3 Hyaluronic-Acid-Based 3D Printing for Biomedical Application

3.3.1 Cell and Organoid Culture

Bio-ink is the soul of 3D printing technology. The selected bio-ink must not only meet the physical and chemical requirements of printing, but also provide an ideal environment for encapsulating cells. However, these requirements are often mutually exclusive. For example, high-concentration bio-ink helps to ensure the stability of the structure during the 3D printing process, but the resulting high-density polymer network may hinder basic cell activities. Ouyang et al. [43] solved this paradox by using complementary network bio-inks. The heat-sensitive gelatin network provides excellent stability during the 3D printing process, while the photocross-linkable hyaluronic acid network allowing for valence cross-linking is used to stabilize the printed structure, and then the 3D printed scaffold can better perform cell encapsulation by thermally dissociating the gelatin network.

Glioblastoma multiforme (GBM) is the deadliest primary brain tumor, and its heterogeneity mainly comes from cellular components and extracellular matrix. In order to further study the potential relationship between them, Tang et al. [44] used the cells and hyaluronic acid derivatives from patients to adjust the 3D printing GBM parameters. The stiffness of the three different regions in the model was established as a three-region bionic model with GBM matrix, brain parenchymal regions in pathological conditions, and brain parenchymal regions in healthy conditions. They found that the rigid ECM microenvironment induced the mesenchymal phenotype associated with tumor recurrence, while the soft ECM microenvironment promoted the rapid proliferation of cells and supported the expansion of cells with classic phenotypes. This 3D printing three-region bionic system provides an effective platform for studying the occurrence and development of GBM. In addition, Tang et al. [45] used hyaluronic acid hydrogel, based on the feature that macrophages and microglia occupy a large part of the tumor volume in progressive or recurrent glioblastoma, fabricated a 3D printing platform constructed by glioblastoma stem cells, macrophages, and astrocytes for further studying the effect of macrophages/microglia on the growth of glioblastoma. They found that in the presence of macrophages and microglia, the proliferation and differentiation of glioblastoma stem cells were more obvious, which also confirmed the effectiveness of this platform in studying multicellular interactions (Figure 3.7).

3.3.2 Tissue Regeneration

For the existing bioprinting technology, creating a 3D structure with interconnected microchannels for the interaction of nutrients and metabolic waste is still a big challenge. Zhou et al. [46] used GelMA, butyramide linked hyaluronic acid, and photoinitiator to form bio-ink. A new method of printing functional living skin based on digital light processing 3D printing technology is presented. This kind of bio-ink is similar to the natural extracellular matrix structure and can quickly gel. Under UV light induction, butyramide-linked hyaluronic acid can react with the amine group on the GelMA chain to adjust the overall mechanical properties. In addition, the 3D printing material can be closely connected to the tissue through the Schiff base reaction. At the same time, they also designed a double-layer 3D printing structure based on the full-layer structure of the skin. The upper simulated epidermal layer has a dense structure to protect the wound from external mechanical shocks and pressure. The lower layer simulates the dermis through the cantilever structure,

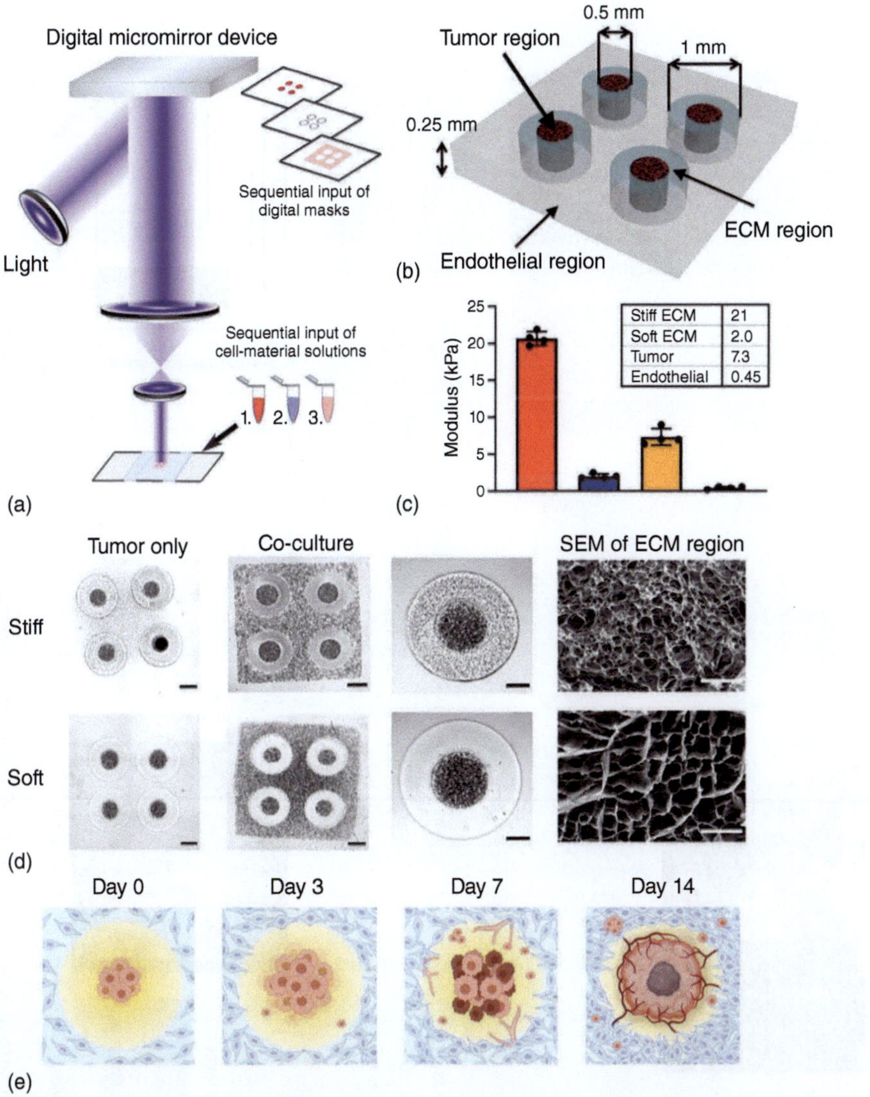

Figure 3.7 (a) Multi-step digital photoprocessing bioprinting method to prepare glioblastoma (GBM) models with biophysical properties of different regions. (b) Dimensions of the models. (c) Stiffness of each region. (d) Images of the rigid and soft GBM models showing different regions. (e) Schematic diagram of GBM formation. Source: Ref. [45]/John Wiley & Sons.

which promotes the perfusion of oxygen/nutrients, the excretion of metabolic waste, and promotes the growth of tissues, thereby promoting wound healing (Figure 3.8).

Hyaluronic acid is also an important part of cartilage tissue. Through 3D printing technology, a well-organized cell carrier can be constructed and applied to cartilage tissue repair. The biggest limitation of the application of hyaluronic acid in the field

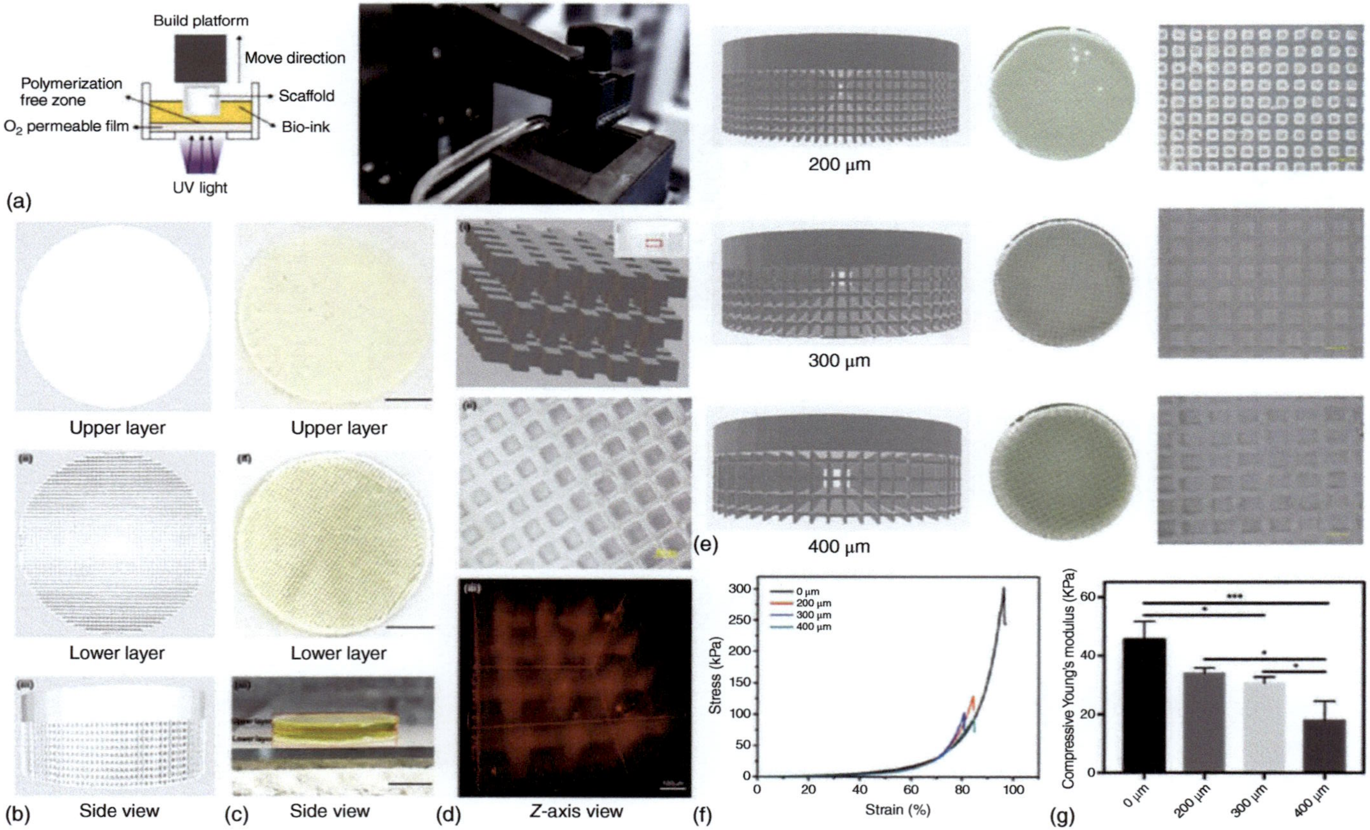

Figure 3.8 (a) A DLP-based 3D printing platform is described. (b) A CAD model of bi-layered skin simulating dermis and epidermis is designed. (c) The designed bi-layered skin structure is printed using DLP. (d) The microstructure of the printed material is observed. (e) Micro-channels of different sizes are designed and printed. (f) The compression properties of the materials of the different sizes of the micro-channels are tested. (g) The elastic modulus of the materials of the different sizes of the micro-channels is measured. Source: Ref. [46]/with permission of Elsevier.

of 3D printing tissue engineering is that its mechanical properties cannot meet the needs of three-dimensional extrusion bioprinting. Antich et al. [47] added alginate to the hyaluronic acid system to give it the ability to quickly cross-link in the presence of calcium ions. The bio-ink both meets the mechanical requirements of 3D printing and retain the unique function of hyaluronic acid to promote cartilage formation and producing matrix components.

References

1. Schaefer, L. and Schaefer, R.M. (2010). Proteoglycans: from structural compounds to signaling molecules. *Cell Tissue Res.* 339: 237–246.
2. Stecco, C., Stern, R., Porzionato, A. et al. (2011). Hyaluronan within fascia in the etiology of myofascial pain. *Surg. Radiol. Anat.* 33: 891–896.
3. Gupta, R.C., Lall, R., Srivastava, A., and Sinha, A. (2019). Hyaluronic acid: molecular mechanisms and therapeutic trajectory. *Front. Vet. Sci.* 6: 192.
4. Schulz, T., Schumacher, U., and Prehm, P. (2007). Hyaluronan export by the ABC transporter MRP5 and its modulation by intracellular cGMP. *J. Biol. Chem.* 282: 20999–21004.
5. Jiang, D.H., Liang, J.R., and Noble, P.W. (2011). Hyaluronan as an immune regulator in human diseases. *Physiol. Rev.* 91: 221–264.
6. Sironen, R.K., Tammi, M., Tammi, R. et al. (2011). Hyaluronan in human malignancies. *Exp. Cell Res.* 317: 383–391.
7. Sneath, R.J.S. and Mangham, D.C. (1998). The normal structure and function of CD44 and its role in neoplasia. *Mol. Pathol.* 51: 191–200.
8. Fallacara, A., Baldini, E., Manfredini, S., and Vertuani, S. (2018). Hyaluronic acid in the third millennium. *Polymers (Basel)* 10: 701.
9. Miguel, S.P., Simoes, D., Moreira, A.F. et al. (2019). Production and characterization of electrospun silk fibroin based asymmetric membranes for wound dressing applications. *Int. J. Biol. Macromol.* 121: 524–535.
10. Cosgrove, B.D., Mui, K.L., Driscoll, T.P. et al. (2016). N-cadherin adhesive interactions modulate matrix mechanosensing and fate commitment of mesenchymal stem cells. *Nat. Mater.* 15: 1297–1306.
11. Vega, S.L., Kwon, M.Y., Song, K.H. et al. (2018). Combinatorial hydrogels with biochemical gradients for screening 3D cellular microenvironments. *Nat. Commun.* 9: 614.
12. Tam, R.Y., Smith, L.J., and Shoichet, M.S. (2017). Engineering cellular microenvironments with photo- and enzymatically responsive hydrogels: toward biomimetic 3D cell culture models. *Acc. Chem. Res.* 50: 703–713.
13. Khetan, S., Guvendiren, M., Legant, W.R. et al. (2013). Degradation-mediated cellular traction directs stem cell fate in covalently crosslinked three-dimensional hydrogels. *Nat. Mater.* 12: 458–465.
14. Yu, S., Duan, Y.Y., Zuo, X.G. et al. (2018). Mediating the invasion of smooth muscle cells into a cell-responsive hydrogel under the existence of immune cells. *Biomaterials* 180: 193–205.

15 Wieduwild, R. and Howarth, M. (2018). Assembling and decorating hyaluronan hydrogels with twin protein superglues to mimic cell-cell interactions. *Biomaterials* 180: 253–264.

16 Hsu, H.H., Ko, P.L., Wu, H.M. et al. (2021). Study 3D endothelial cell network formation under various oxygen microenvironment and hydrogel composition combinations using upside-down microfluidic devices. *Small* 17: e2006091.

17 Baker, A.E.G., Bahlmann, L.C., Tam, R.Y. et al. (2019). Benchmarking to the gold standard: hyaluronan-oxime hydrogels recapitulate xenograft models with in vitro breast cancer spheroid culture. *Adv. Mater.* 31: e1901166.

18 Wu, J.P.J., Cheng, B., Roffler, S.R. et al. (2016). Reloadable multidrug capturing delivery system for targeted ischemic disease treatment. *Sci. Transl. Med.* 8: 365ra160.

19 Shamskhou, E.A., Kratochvil, M.J., Orcholski, M.E. et al. (2019). Hydrogel-based delivery of Il-10 improves treatment of bleomycin-induced lung fibrosis in mice. *Biomaterials* 203: 52–62.

20 MacArthur, J.W., Purcell, B.P., Shudo, Y. et al. (2013). Sustained release of engineered stromal cell-derived factor 1-α from injectable hydrogels effectively recruits endothelial progenitor cells and preserves ventricular function after myocardial infarction. *Circulation* 128: S79–S86.

21 Wang, L.L., Liu, Y., Chung, J.J. et al. (2017). Local and sustained miRNA delivery from an injectable hydrogel promotes cardiomyocyte proliferation and functional regeneration after ischemic injury. *Nat. Biomed. Eng.* 1: 983–992.

22 Saleh, B., Dhaliwal, H.K., Portillo-Lara, R. et al. (2019). Local immunomodulation using an adhesive hydrogel loaded with miRNA-laden nanoparticles promotes wound healing. *Small* 15: 1902232.

23 Zhang, K.Y., Jia, Z.F., Yang, B.G. et al. (2018). Adaptable hydrogels mediate cofactor-assisted activation of biomarker-responsive drug delivery via positive feedback for enhanced tissue regeneration. *Adv. Sci.* 5: 1800875.

24 Zhang, L.C., Xiang, Y., Zhang, H.B. et al. (2020). A biomimetic 3D-self-forming approach for microvascular scaffolds. *Adv. Sci.* 7: 1903553.

25 Kang, B., Shin, J., Park, H.J. et al. (2018). High-resolution acoustophoretic 3D cell patterning to construct functional collateral cylindroids for ischemia therapy. *Nat. Commun.* 9: 5402.

26 Mothe, A.J., Tam, R.Y., Zahir, T. et al. (2013). Repair of the injured spinal cord by transplantation of neural stem cells in a hyaluronan-based hydrogel. *Biomaterials* 34: 3775–3783.

27 Tay, A., Sohrabi, A., Poole, K. et al. (2018). A 3D magnetic hyaluronic acid hydrogel for magnetomechanical neuromodulation of primary dorsal root ganglion neurons. *Adv. Mater.* 30: e1800927.

28 Chae, J.J., Jung, J.H., Zhu, W. et al. (2021). Drug-free, nonsurgical reduction of intraocular pressure for four months after suprachoroidal injection of hyaluronic acid hydrogel. *Adv. Sci.* 8: 2001908.

29 Raia, N.R., Jia, D., Ghezzi, C.E. et al. (2020). Characterization of silk-hyaluronic acid composite hydrogels towards vitreous humor substitutes. *Biomaterials* 233: 119729.

30 Koivusalo, L., Kauppila, M., Samanta, S. et al. (2019). Tissue adhesive hyaluronic acid hydrogels for sutureless stem cell delivery and regeneration of corneal epithelium and stroma. *Biomaterials* 225: 119516.

31 Wang, S.Y., Kim, H., Kwak, G. et al. (2018). Development of biocompatible HA hydrogels embedded with a new synthetic peptide promoting cellular migration for advanced wound care management. *Adv. Sci.* 5: 1800852.

32 Wu, J., Chen, A.Q., Zhou, Y.J. et al. (2019). Novel H_2S-releasing hydrogel for wound repair via *in situ* polarization of M2 macrophages. *Biomaterials* 222: 119398.

33 El-Aassar, M.R., Ibrahim, O.M., Fouda, M.M.G. et al. (2020). Wound healing of nanofiber comprising polygalacturonic/hyaluronic acid embedded silver nanoparticles: in-vitro and in-vivo studies. *Carbohydr. Polym.* 238: 116175.

34 Séon-Lutz, M., Couffin, A.C., Vignoud, S. et al. (2019). Electrospinning in water and in situ crosslinking of hyaluronic acid/cyclodextrin nanofibers: towards wound dressing with controlled drug release. *Carbohydr. Polym.* 207: 276–287.

35 Hussein, Y., El-Fakharany, E.M., Kamoun, E.A. et al. (2020). Electrospun PVA/hyaluronic acid/L-arginine nanofibers for wound healing applications: nanofibers optimization and in vitro bioevaluation. *Int. J. Biol. Macromol.* 164: 667–676.

36 Movahedi, M., Asefnejad, A., Rafienia, M., and Khorasani, M.T. (2020). Potential of novel electrospun core-shell structured polyurethane/starch (hyaluronic acid) nanofibers for skin tissue engineering: in vitro and in vivo evaluation. *Int. J. Biol. Macromol.* 146: 627–637.

37 Rao, S.S., Nelson, M.T., Xue, R.P. et al. (2013). Mimicking white matter tract topography using core-shell electrospun nanofibers to examine migration of malignant brain tumors. *Biomaterials* 34: 5181–5190.

38 Li, X.W., Cho, B., Martin, R. et al. (2019). Nanofiber-hydrogel composite-mediated angiogenesis for soft tissue reconstruction. *Sci. Transl. Med.* 11: eaau6210.

39 Song, K.H., Heo, S.J., Peredo, A.P. et al. (2020). Influence of fiber stiffness on meniscal cell migration into dense fibrous networks. *Adv. Healthc. Mater.* 9: 1901228.

40 Kim, I.L., Khetan, S., Baker, B.M. et al. (2013). Fibrous hyaluronic acid hydrogels that direct MSC chondrogenesis through mechanical and adhesive cues. *Biomaterials* 34: 5571–5580.

41 Li, M., Zhang, X.L., Jia, W.B. et al. (2019). Improving in vitro biocompatibility on biomimetic mineralized collagen bone materials modified with hyaluronic acid oligosaccharide. *Mater. Sci. Eng. C* 104: 110008.

42 Apsite, I., Constante, G., Dulle, M. et al. (2020). 4D biofabrication of fibrous artificial nerve graft for neuron regeneration. *Biofabrication* 12: 035207.

43 Ouyang, L.L., Armstrong, J.P.K., Lin, Y.Y. et al. (2020). Expanding and optimizing 3D bioprinting capabilities using complementary network bioinks. *Sci. Adv.* 6: abc5529.

44 Tang, M., Xie, Q., Gimple, R.C. et al. (2020). Three-dimensional bioprinted glioblastoma microenvironments model cellular dependencies and immune interactions. *Cell Res.* 30: 833–853.

45 Tang, M., Tiwari, S.K., Agrawal, K. et al. (2021). Rapid 3D bioprinting of glioblastoma model mimicking native biophysical heterogeneity. *Small* 17: 2006050.

46 Zhou, F.F., Hong, Y., Liang, R.J. et al. (2020). Rapid printing of bio-inspired 3D tissue constructs for skin regeneration. *Biomaterials* 258: 120287.

47 Antich, C., de Vicente, J., Jiménez, G. et al. (2020). Bio-inspired hydrogel composed of hyaluronic acid and alginate as a potential bioink for 3D bioprinting of articular cartilage engineering constructs. *Acta Biomater.* 106: 114–123.

4

Sources, Structures, and Properties of Chitosan

Chitosan is the second largest natural polysaccharide after cellulose [1], and it is also the only cationic natural linear polysaccharide. Chitosan is mainly derived from animal carapace. In recent years, some researchers have successfully prepared chitosan from fungus [2].

The development history of chitosan can be traced back to the nineteenth century. Rouget first discussed the acetylation form of chitosan in 1859, and then it was widely used in biomedical field because of its good biocompatibility, biodegradability, antibacterial ability, and porous properties. Chitosan is composed of glucosamine and *N*-acetyl glucosamine units linked by β-(1-4) glycosidic bonds, in which the content of glucosamine is also known as deacetylated. According to the different material sources and extraction methods, the molecular weight and degree of deacetylation of chitosan are also very different [3], and the content of amino group can be effectively controlled by adjusting the degree of deacetylation of chitosan [4].

Chitosan contains three functional groups, namely amino group, primary hydroxyl group, and secondary hydroxyl group, and the properties of amino groups are the starting point of most material designs at present. Amino groups endow chitosan with cationic properties, which enables chitosan to react with glycosaminoglycans and proteoglycans distributed throughout the human body to enhance the adhesion between chitosan and tissue proteins. Studies have shown that chitosan can open the mucosal barrier channel to facilitate drug transport [5]. In addition, the PKA value of chitosan is about 6.5, and its structural network will swell in acidic environment and contract in neutral and alkaline environments [6]. The hydrogel derived from chitosan has excellent mechanical properties, and its porous structure is conducive to material exchange and cell adhesion, so it is also widely used in the field of tissue engineering.

4.1 Chitosan-Based Hydrogel for Biomedical Application

4.1.1 Cell and Organoid Culture

Blood vessels are an important channel for the exchange of substances in the human body. In addition to the function of transmitting signals to maintain the integrity

Natural Polymers for Biomedical Applications, First Edition. Wenguo Cui and Lei Xiang.

of various biological functions, nerves also play a role in nourishing tissue cells. Therefore, the regeneration of blood vessels and nerves has attracted much attention in the field of tissue engineering. Through targeted delivery of stem cells, promoting angiogenesis from the paracrine pathway is a promising treatment method. The therapeutic effect mainly depends on whether the stem cells can be delivered to the injury site to the greatest extent and maintain good biological activity. In lower-limb ischemia models, traditional hydrogels often cause structural damage due to the dynamic load of the lower limbs, resulting in the dissipation of stem cells and adversely affecting the therapeutic effect. Young et al. [7] developed an injectable, in situ gelled hydrogel for the mechanically dynamic intramuscular targeted delivery strategy of adipose-derived stem cells/stromal cells. They used ammonium persulfate and *N*,*N*,*N*′,*N*′-tetramethyl ethylenediamine as initiators, and copolymerized with polytrimethylene carbonate-*b*-polyethylene glycol-*b*-polytrimethyl carbonate. The compound was thermally induced to cross-link the methacrylate glycol chitosan hydrogel. The prepared scaffold has an ultimate compressive strain above 75%, maintains good mechanical properties in a compression fatigue test at a physiological level, and can achieve rapid gelation within three minutes under the condition of a low concentration of initiator. Loaded adipose-derived stem cells show a high survival rate (>90%) after being cultured under normoxia and hypoxia for two weeks. The encapsulated cells produce angiogenesis and release chemotactic cytokines under continuous hypoxic conditions. In the mouse model of hindlimb ischemia, this hydrogel can maintain the integrity of the material for 28 days, and compared with the untreated control group, the blood vessel density of mesenchymal stem cells is significantly increased. Bacterial rhodopsin plasmid has photoelectric properties and can be used as an optical switch for cell activation. Hsieh et al. [8] mixed it with an injectable chitosan hydrogel to efficiently transport the bacterial rhodopsin plasmid into the cell while maintaining the viability of the bacterial rhodopsin plasmid cell. Under the stimulation of green light, bacterial rhodopsin-mediated neural stem cells can proliferate and differentiate into neurons specifically in vitro.

4.1.2 Tissue Regeneration

Intermolecular interaction forces such as hydrogen bonds, electrostatic interactions, and van der Waals forces can give hydrogels adhesive properties. However, on the surface of the wet tissue, the presence of water molecules hinders the intermolecular interaction and thus reduces the adhesion of the hydrogel. Yuk et al. [9] grafted polyacrylic acid and *N*-hydro succinimide ester onto the surface of chitosan to prepare a dry adhesive. This adhesive can enhance the adhesion properties of the hydrogel by removing the surface interface water. Experiments have shown that it can achieve firm adhesion to the wet tissue surface within four seconds. In order to meet the wide application in the field of tissue engineering, how to improve the anti-fatigue performance and self-healing ability of hydrogels has also become the research direction of the majority of researchers. Compared with the fragile mechanical properties of single network hydrogels, ion–covalent cross-linked double-network hybrid hydrogels have attracted more and more attention for their

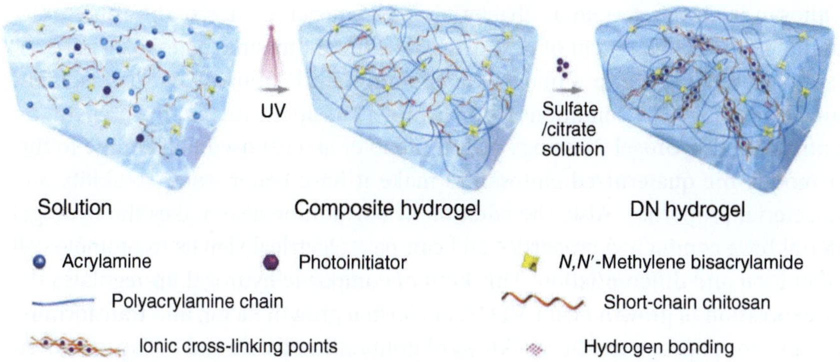

Figure 4.1 Fabrication of hybrid CS-PAM ionic–covalent DN hydrogels. Source: Ref. [10]/John Wiley & Sons.

excellent self-healing ability and fatigue resistance. Yang et al. [10] developed a high-performance chitosan-polyacrylamide ion–covalent double-network hydrogel, and flexibly adjusted its structure and mechanical properties by adjusting the cross-linking time. Short-chain chitosan was first integrated into the covalent polyacrylamide network to prepare a highly ductile composite hydrogel. Then, a multivalent anion solution is used for cross-linking to form a rigid chitosan network to further enhance the mechanical properties of the hydrogel. The dual-network hydrogel has high tensile strength (≈5.6 MPa), strong elastic modulus (≈1.3 MPa), ultra-high fracture toughness (≈14.0 kJ/m), rapid self-healing ability (four hours recovery rate is greater than 90%), and excellent fatigue resistance (after 60 compressions to 50% of the original volume, it can still be restored to 89% of the original value). Linked network structure is another commonly used method to enhance the mechanical properties of hydrogels. Different from the improvement of homogeneity of double networks or the addition of energy dissipation networks, connected network hydrogels cooperate with each other by sharing connection points, so that all composed networks have similar energy dissipation mechanisms to effectively bear the burden of the entire system (Figure 4.1).

Xu et al. [11] electrostatically cross-linked chitosan–gelatin complex with polyvalent sodium phytate to prepare a bio-linked network hydrogel. While achieving high compressive modulus and toughness, this hydrogel also exhibits excellent self-healing and anti-fatigue capabilities, which provides a new idea for the design of biocompatible rigid and tough network hydrogels.

In addition, Chedly et al. [12] prepared a fragmented physical hydrogel suspension using only chitosan and water, and evaluated the damaged spinal cord tissue and pure chitosan-based implantation. They found that the chitosan–FPHS implantation immediately after the bilateral dorsal spinal cord was cut can promote the reconstruction of spinal cord tissue and blood vessels. Animal experiments have shown that the astrocyte protrusions are mainly toward the injury site, making the boundary between the injury site and the intact tissue allow a large number of axons to grow back to the injury site, and the growing axons are myelinated or endogenous wrapped cells promote nerve repair and regeneration.

Chitosan itself has certain antibacterial and hemostatic effects, and it has great potential in the development of new hydrogel wound dressing materials. However, its antibacterial activity in a non-acidic environment is limited, and its insolubility under physiological conditions further hinders its application in the preparation of antibacterial hydrogel dressings in situ. Zhao et al. [13] used polyaniline to further modify the quaternized chitosan to make it have better water solubility and antibacterial properties. Also, the addition of polyaniline also makes the hydrogel material have conductive properties and can pass electrical signals to promote cell proliferation and differentiation. This kind of composite hydrogel up-regulates the gene expression of growth factor VEGF, epidermal growth factor, and transforming growth factor β, promotes the thickness of granulation tissue and collagen deposition, and significantly promotes the wound healing process of the full-thickness skin defect model in vivo. Hot springs can promote human blood circulation and accelerate human metabolism. In addition, the iron, silicon, and other elements contained in the hot springs also promote vascular activity. Sheng et al. [14] were inspired by hot springs and proposed for the first time the concept of "fake hot spring biological materials" for tissue regeneration. A hydrogel for chronic wound healing was prepared using Fe_2SiO_4 bio-ceramics and *N*,*O*-carboxymethyl chitosan as the matrix. The obtained composite hydrogel has good Fe^{2+}/SiO_4/ion release and photothermal performance, providing a hot spring-like environment for chronic wounds. Studies have shown that the simulated hot spring hydrogel combines biologically active Fe^{2+}/SiO_4/ions with a mild thermal environment, and by activating different angiogenic factors and signal pathways, it is very effective in stimulating angiogenesis and promoting wound healing. Wound healing after skin tumor resection faces the risk of slow wound healing and possible residual tumor cells leading to recurrence. Most of the current wound excipients focus on promoting the rapid healing of the wound surface and fail to remove the possible residual tumor cells. Photothermal therapy is also a very effective method for tumor treatment in recent years. Wang et al. [15] integrated titanium dioxide (B-TiO_2-x) nanoparticles into a chitosan matrix to prepare an injectable thermosensitive hydrogel glue. BTiO_2-x nanocrystals have a crystalline/amorphous core–shell structure and abundant oxygen vacancies, making the thermal gel capable of photothermal therapy and photodynamic therapy under single-wavelength near-infrared laser irradiation. In addition, this photothermal gel not only supports the adhesion, proliferation, and migration of normal skin cells, but also exhibits the function of promoting skin tissue regeneration and repair in mouse chronic wound models. The photothermal gel can well solve the wound healing problem after skin tumor resection.

The success of wound healing and tissue defect repair is largely related to the ability to control infection. Li et al. [16] formed a polyvinyl alcohol hydrogel modified by chitosan, polydopamine, and NO release agent on the red phosphorous nanomembrane deposited on the surface of titanium implants. Under near-infrared light, the NO released in the hydrogel reacts with the superoxide ($\cdot O^{2-}$) produced by the red phosphorous nano-film to form peroxynitrite ($\cdot ONOO^-$), which can effectively remove methicillin-resistant gold *Staphylococcus aureus* biofilm. In the absence of light, the inherent antibacterial ability of chitosan can also inhibit the formation of biofilms. Also based on the concept of light-triggered drug release, Gao et al. [17] mixed ciprofloxacin-loaded polydopamine nanoparticles with ethylene glycol chitosan to make an injectable hydrogel, which was positively charged by chitosan. Adhesive bacteria with dopamine nanoparticles can be easily adsorbed on the surface of the hydrogel, and ciprofloxacin can be released on demand to kill bacteria under the irradiation of near-infrared light.

After myocardial infarction, normal myocardial cells in the lesion area are spontaneously repaired and replaced by scar tissue due to ischemic necrosis. However, scar tissue cannot effectively conduct electrical signals and eventually causes ventricular dysfunction. The restoration of normal electrical signal conduction plays a vital role in the recovery of overall cardiac function after infarction. Mihic et al. [18] We grafted pyrrole onto the clinically tested biomaterial chitosan to form a polypyrrole–chitosan hydrogel, thereby forming a conductive polymer. Compared with traditional chitosan hydrogels, polypyrrole–chitosan can enhance calcium signal conduction in neonatal rat cardiomyocytes and improve the electrical coupling between skeletal muscles 25 mm apart. Injecting polypyrrole–chitosan one week after myocardial infarction can shorten the QRS interval, increase the lateral activation speed, and suggest that the electrical conduction is improved. Echocardiography and pressure–volume analysis showed that after eight weeks of injection, various cardiac activity indicators including ejection fraction were significantly improved. In addition, He Sheng et al. [19] also further studied the electrical conductivity of polypyrrole–chitosan hydrogel and its role in cardio-protection. They found that by injecting polypyrrole–chitosan hydrogel into the infarct area, the electrical resistance of the tissue can be reduced by 30% and the heart contraction was resynchronized.

Hydrogel is a hot material in the field of cartilage tissue engineering. Man et al. [20] used chitosan hydrogel-decalcified bone matrix hybrid scaffold for transplantation of allogeneic chondrocytes. They found that chitosan hydrogel can fill the large pores in the demineralized bone matrix, thereby improving cell adhesion and distribution. In the rabbit cartilage defect model, the expression of BMP-7, HGF, IGF-1, and other genes that are conducive to cartilage regeneration was up-regulated one month after implantation of the hybrid stent. After 24 weeks, the defect was found to be completely filled, which confirmed that it was in the cartilage. Similar to cartilage tissue repair, bone repair involves multiple factors such as extracellular matrix, various inducing factors, and cell–cell interactions. In order to meet the influence of multiple factors in the bone repair process at the same time, Lee et al. [21] prepared a new type of caffeic-acid-modified chitosan and nanosilicate through a variety of covalent and non-covalent cross-linking. The catechol group in the hydrogel component enhances the mechanical properties of the hydrogel and also enables the hydrogel to be injected into defects of various shapes. The phenolic hydroxyl group rich in caffeic acid has antibacterial and antioxidant activities. In addition, nano-clay can increase osteoinductive signals through the Wnt/b-catenin pathway to promote bone regeneration. Also using nano-clay particles, Cui et al. [22] successfully prepared nano-chitosan hydrogels with microporous interconnected

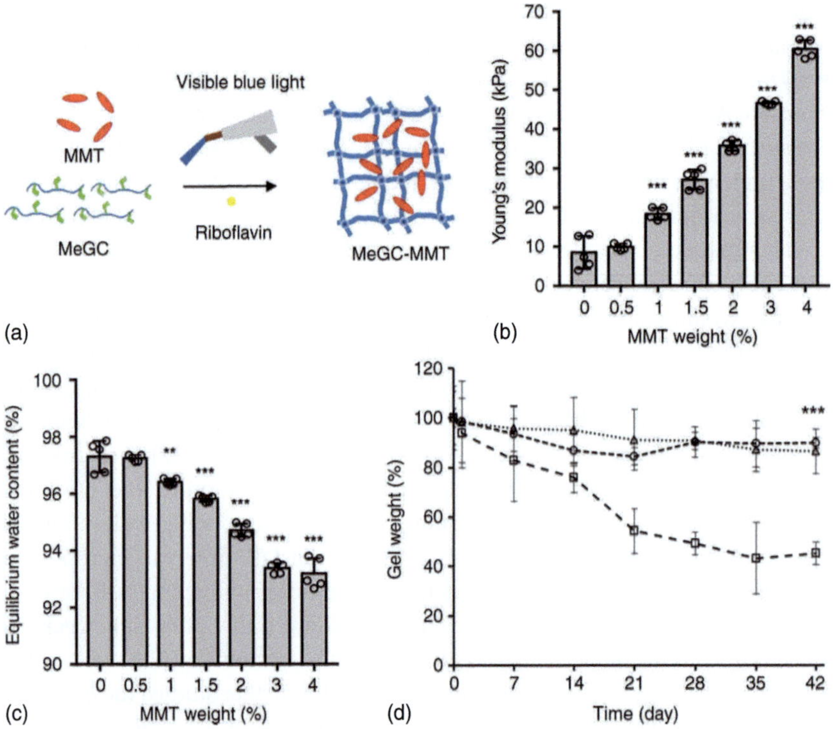

Figure 4.2 (a) Curing of MeGC-MMT nanocomposite hydrogels by visible blue light cross-linking using riboflavin as photoinitiator. (b) Young's modulus of MeGC-MMT hydrogels with different MMT contents. (c) Equilibrium water content of MeGC-MMT hydrogels. (d) Degradation curves of MeGC-MMT hydrogels with different MMT contents in PBS. Source: Ref. [22]/Springer Nature/CC BY 4.0.

microporous structure by introducing montmorillonite nanoparticles with intercalation chemistry. The addition of montmorillonite enhanced Young's modulus of the overall hydrogel and slowed down the degradation rate of the hydrogel. In vitro experiments confirmed that this hydrogel can promote the proliferation and attachment of encapsulated mesenchymal stem cells and induce their differentiation. In the mouse skull defect model, without adding additional growth factors and stem cells, chitosan montmorillonite hydrogel can recruit natural cells and promote skull healing, which confirms their potential in the field of tissue engineering (Figure 4.2).

Rheumatoid arthritis can cause severe bone destruction, but current treatments mostly focus on anti-inflammatory development and neglect bone repair. Pan et al. [23] combined black phosphorous nanosheets with platelet-rich plasma–chitosan thermosensitive hydrogel to develop a new drug delivery system. Black phosphorous nanosheets can generate local heat under near-infrared radiation, and at the same time deliver active oxygen to the inflamed joints to remove the proliferated synovial tissue and improve joint movement. Injectable chitosan thermosensitive hydrogel can control the release of black scale degradation products and provide abundant raw materials for osteogenesis. In addition, platelet-rich plasma can effectively improve the adhesion of mesenchymal stem cells on the chitosan thermosensitive hydrogel and promote their osteogenic differentiation. This temperature-sensitive hydrogel can significantly improve the degree of joint swelling in arthritic mice, confirming the beneficial effect of promoting osteogenesis on rheumatoid arthritis.

4.2 Chitosan-Based Electrospinning for Biomedical Application

4.2.1 Drug and Cell Delivery

In terms of drug delivery, electrospinning can effectively control the release process of various hydrophilic drugs. Saudi et al. [24] synthesized polycaprolactone–chitosan nano-network–nanofiber electrospinning network to optimize the release of diclofenac sodium (a hydrophilic anti-inflammatory drug) process. Through material characterization, they found that the presence of diclofenac sodium is the key to the formation of electrospun nanonetworks. In addition, the prepared electrospun fiber can improve the adhesion, vitality and proliferation ability of 3T3 cells in the nanonetwork–nanofiber network by controlling diclofenac sodium. Electrospinning loaded with diclofenac sodium can show a stable release profile in an experiment for up to 14 days, thereby improving the anti-inflammatory effect of the drug. Polyester polymers are often used as matrix materials for various tissue engineering. However, they will inevitably produce acidic degradation products in the body and cause aseptic inflammation.

Chitosan has unique acid neutralization properties, which can fundamentally solve the inflammation caused by acid degradation products. Shen et al. [25] solved the above problems by coating a layer of alkaline chitosan on the surface of PLGA fibers by coaxial electrospinning. In experiments on human dermal fibroblasts in a simulated acidic environment, the acid-neutralizing ability of chitosan can significantly reduce the secretion of inflammatory factors and down-regulate the expression of related inflammatory genes. In vitro biocompatibility evaluation showed that although the cell adhesion ability of chitosan/PLGA fibers is worse than that of PLGA fibers, the cell compatibility is good, which promotes cell migration and collagen secretion. In addition, in vivo subcutaneous implantation experiments showed that chitosan/PLGA fibers significantly reduced the recruitment of inflammatory cells and the formation of foreign body giant cells. This pH neutral fiber can reduce the possible side effects of graft degradation on the basis of promoting cell activity, and this highly oriented fiber has great application value in tissue engineering of anisotropic structures such as tendons. The construction of highly oriented electrospun fibers is of great significance for the repair and regeneration of tendons, but the unique structure of electrospun fibers alone is slightly weak in promoting tendon regeneration. Zhang et al. [26] further studied the role of directed electrospinning in regulating stem cell differentiation. They used a stable jet electrospinning method to prepare nanofibers that can mimic the microstructure and mechanical properties of tendons. Using chitosan as a matrix material can effectively prevent tissue adhesion during implantation. They first differentiated HiPSCs into MSCs on a smooth plastic surface and then planted the mesenchymal stem cells on a fiber scaffold. The results show that this highly oriented nanofiber can promote the differentiation of mesenchymal stem cells into the tendon lineage. After culturing on the arranged fiber scaffold, the expression of all integrin subunits and non-muscle myosin II B increased, indicating that the structure of the fiber may regulate the remodeling of the cytoskeleton through the integrin signaling pathway. In addition, the aligned fibers produced a higher expression of tendon-specific transcription factors (SCX, MKX) and tendon-related extracellular matrix (ECM) molecules. Their research provides strong evidence for artificially regulating the directed differentiation of stem cells.

Stem cell therapy not only has great application value in promoting tendon regeneration, stem-cell-based cardiac patch has also become a potential regeneration strategy for the treatment of myocardial infarction. Chen et al. [27] electrospun modified cellulose nanofibers with chitosan/silk fibroin multilayer films and prepared electrospun fiber patches by layer-by-layer coating (LBL) technology. Adipose-tissue-derived mesenchymal stem cells were planted on the patch and then implanted on the epicardium of the rat myocardial infarction area. Thanks to the LBL technology, silk fibroin with good elasticity and biodegradability and chitosan with high bonding strength and antibacterial properties can be combined. The experimental results showed that the survival rate of stem cells in the cell nano-patch group was higher than that of the intramyocardial injection group, the ventricular remodeling of the nano-patch group was reduced, and the left ventricular end-diastolic volume and end-systolic volume were smaller than those of the control group. In addition, compared with the biomarkers of myocardial fibrosis (transforming growth factor-β1, P-Smad3, and Smad3) and the biomarkers of ventricular remodeling (BNP, β-MHC/α-MHC ratio) in the cell nano-patch treatment expression in the heart is reduced.

Also using the principle that the core–shell structure can impart different characteristics to electrospun fibers, Shalumon et al. [28] further studied the role of electrospun fibers with a core–shell structure in inducing bone formation. They use silk fibroin/chitosan/nano-hydroxyapatite/bone morphogenetic protein-2 as the core of the fiber, and a shell composed of silk fibroin/chitosan/nano-hydroxyapatite is used to control bone morphology the release of protein-2 occurs. By adjusting the thickness of the shell, researchers can easily control the release time of BMP-2 to induce osteogenic differentiation of stem cells (Figure 4.3).

4.2.2 Tissue Regeneration

Based on the LBL preparation strategy, the applications of electrospun fibers in the biomedical field have also been significantly expanded. Huang et al. [29] reported a method of polycaprolactone/cellulose acetate co-electrospinning and electrospinning modified by LBL self-assembly of positively charged chitosan and negatively charged type I collagen. Experiments have proved that the biomimetic nanofiber matrix of the LBL structure may promote cell migration by up-regulating

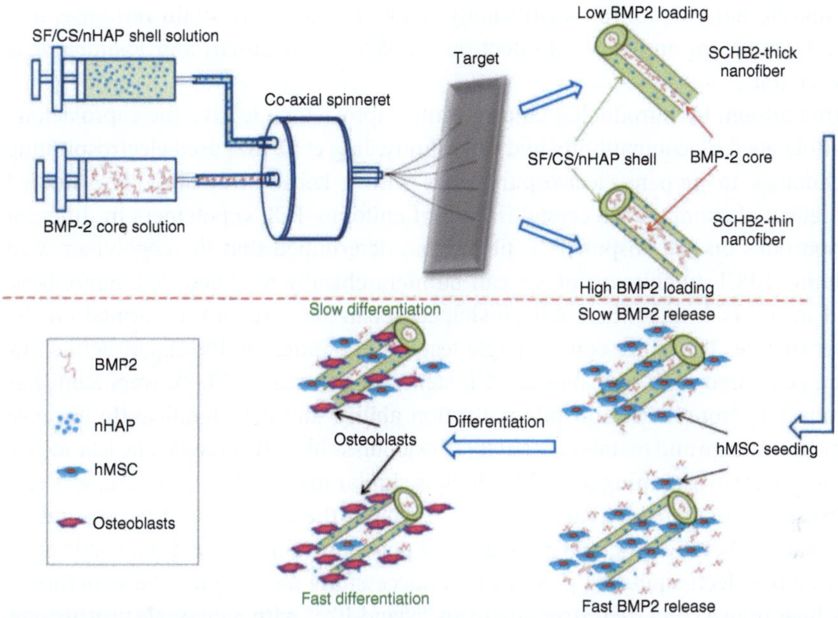

Figure 4.3 Schematic representation of the preparation of SCHB2-thick and SCHB2-thin NFMs through coaxial electrospinning and their influence on hMSCs. Source: Ref. [28]/ American Chemical Society.

the secretion of ECM protein and triggering the integrin/FAK signaling pathway, thereby promoting skin regeneration, thus proving that the nanofiber matrix may be used to quickly restore the structure of wounded skin and functional characteristics. Bazmandeh et al. [30], in order to simulate the fiber/gel structure of the skin ECM, used biopolymers similar in structure to ECM, namely chitosan, gelatin, and hyaluronic acid, and prepared a bionic multi-component scaffold. They used dual electrospinning technology to simultaneously prepare chitosan–gelatin and chitosan–hyaluronic acid composite nanofibers. When the scaffold is exposed to water, the chitosan–hyaluronic acid fiber will form a gel state, while the chitosan–gelatin fiber retains its fiber structure, forming a fiber–gel hybrid structure. This composite structure further promotes cell proliferation and adhesion on the basis of maintaining the original mechanical stability, and animal experiments have proved its strong ability to promote wound healing, and it can form a new structure similar to the natural skin structure. The repair of wounds in some special areas (wounds after hernia repair) has high requirements for the mechanical properties and antibacterial properties of wound accessories. However, most of the currently used wound excipients can only meet one condition. Keirouz et al. [31] used shaft electrospinning technology to prepare shell–core nanofibers with dual drug release effects and mechanical strength to meet clinical needs. The nylon-6 in the core enhances the mechanical properties of the fiber, and the chitosan/polyoxyethylene in the shell provides continuous antibacterial effect. Compared with the corresponding single polymer electrospun mesh, the composite nanowire mesh significantly improves the stress–strain response, and has a significant antibacterial effect on *Staphylococcus aureus* and *Pseudomonas aeruginosa*.

In addition, by introducing chitosan into caprolactone to give the caprolactone scaffold good biocompatibility and cell affinity, Jing et al. [32] used electrospinning technology to prepare electrospun fibers with a kebab structure. They studied the effect of spontaneous crystallization of chitosan–PCL copolymers in different proportions on electrospun PCL fibers, and determined that the copolymer with a ratio of PCL to chitosan of 8.8 can be hierarchically modified PCL nanofibers, and in the PCL nanofibers A well-shaped kebab-like structure is formed on the fiber surface. The water contact angle test and the imitation life experiment show that, compared with the smooth PCL stent, the chitosan–PCL skewers stent has stronger hydrophilicity and mineralization ability, and can significantly improve the survival rate and metabolic activity of osteoblast-like MG63 cells. The island-like structure electrospinning scaffold, which is similar to the kebab structure, was also successfully prepared by Xu et al. [33]. Through the automatic phase separation mechanism between the two incompatible polymers (chitosan and polylactic acid) during the electrospinning process, they successfully transformed the structure of the fiber from a core–shell structure to an "island-like" with nano-scale protrusions. The size and shape of the island-shaped protrusions can be effectively controlled by controlling the evaporation rate of the solvent and the crystallization process of PLA. Experiments have shown that electrospinning with island-like protrusions is more conducive to cell adhesion, while the chitosan component can provide biological information guidance for cells (for example, enhance the alkaline phosphatase activity of MC3T3-E1 cells and promote surface mineralization). By controlling the size and shape of the island structure, the hydrophilic core and hydrophobicity of the fiber can be further balanced (Figure 4.4).

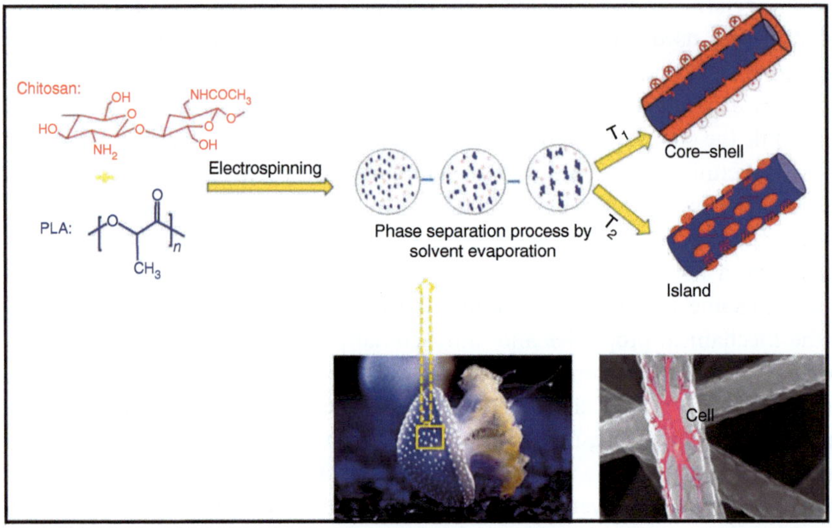

Figure 4.4 Preparation process of CS islanded-structured scaffolds. Source: Ref. [33]/ American Chemical Society.

4.3 Chitosan-Based 3D Printing for Biomedical Application

4.3.1 Cell Behavior Regulation

Elviri et al. [34] confirmed in a previous study that a simple 3D printing structure with an opening of 400 μm was proven to significantly improve the adhesion and proliferation of human fibroblasts. In a follow-up study [35] they prepared 3D printed scaffolds with or without chitosan film as the base layer with an opening of 200 μm, and for the first time monitored the individual culture and co-culture of fibroblasts and keratinocytes. In order to deepen the understanding of the role of 3D in accurately improving the results of in vitro experiments, the experimental results show that the dense chitosan base can effectively avoid cell shedding, extend the adhesion time of cells on the scaffold, and promote the interconnection between cells to form a continuous layered structure, which has great potential in promoting skin integrity repair potential.

4.3.2 Drug Delivery

In addition to the various conventional structure 3D bio-scaffolds that can be prepared by 3D printing technology, the microfluidizers prepared by this technology also have a wide range of applications in drug delivery. At present, most of the drug delivery of microfluidizers is through passive diffusion, which is difficult to meet the requirements of on-demand drug delivery. Bozuyuk et al. [36] proposed a two-photon 3D printing technique to prepare a magnetically driven double-helix microswimming device. They first synthesized a photosensitive form of chitosan methacrylate. The amino group on the micro swimmer was modified by doxorubicin through a photocleavable linker. The release of doxorubicin can be adjusted by controlling the light to satisfy the drug time positioning of release. At the same time, they added iron oxide nanoparticles to the bio-ink to enable the microswimming device to sense the changes in the magnetic field and to satisfy the spatial positioning of the drug release. This method of realizing the dual positioning of time and space for drug release by controlling the magnetic field and illumination opens up new ideas for drug delivery (Figure 4.5).

4.3.3 Tissue Regeneration

The development of biological inks based on biological materials is essential for the application of 3D printing technology in the field of tissue engineering. How to design and prepare bio-inks with good flow properties, mechanical properties, and other properties that can promote cell or tissue regeneration (such as conductivity) is the key to expanding the application of 3D printing technology in the biomedical field.

The conduction of electrical signals has been proven to have great potential in promoting cell proliferation and adhesion. Therefore, electrically responsive three-dimensional scaffolds are a hot research field in tissue engineering. The electrical and mechanical properties of hydrogel materials prepared from pure

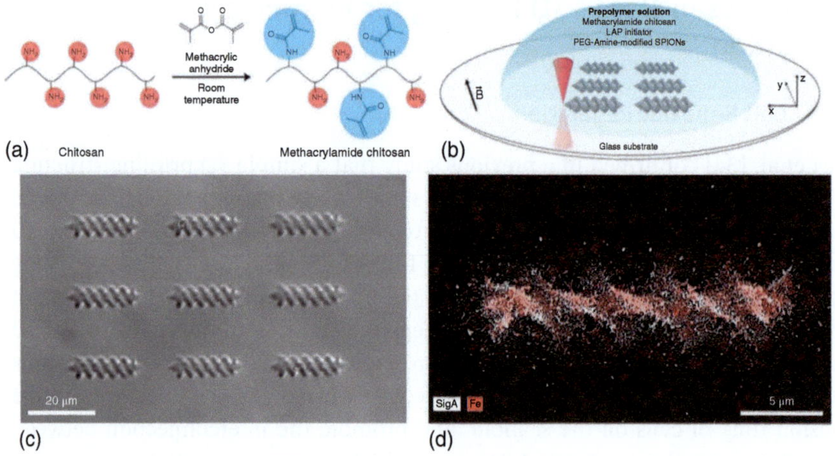

Figure 4.5 (a) Photocross-linkable methacryloyl chitosan was synthesized. (b) Three-dimensional printing of microfluidic channels using two-photon direct laser writing technique. (c) Optical photographs of printed microfluidic channels in a 3 × 3 array. (d) Energy spectroscopy showed the presence of iron in the microfluidic channels. Source: Ref. [36]/American Chemical Society.

chitosan methacrylate cannot meet the needs of cell growth and 3D printing. Sayyar et al. [37] used graphene oxide as a filler. The UV-cross-linked chitosan methacrylate-conductive hydrogel was prepared. Graphene is uniformly dispersed in the polymer matrix and will not affect the UV cross-linking process of the hydrogel. Compared with chitosan–lactic acid–graphene hydrogel samples, the mechanical properties of graphene oxide–methacrylate chitosan have been improved, which can meet the needs of preparing multilayer scaffolds through 3D printing technology. Movement is eternal, while stillness is relative. In actual application scenarios of tissue engineering scaffolds, various mechanical stresses or movements will adversely affect the integrity of the 3D printed scaffold. Therefore, the research and development of a bio-ink with self-healing ability has become the focus of many researchers. Liu et al. [38] used phenol-functionalized chitosan and dibenzaldehyde-terminated telescrew polyethylene glycol as raw materials to develop a self-healing chitosan for injectable hydrogels and printable inks. The introduced phenol groups make the hydrogel have fast gelation ability and good self-healing ability, can maintain stable rheological properties during printing, and further stabilize its structure through secondary cross-linking after printing. This kind of bio-ink can be printed separately or assembled by components to prepare various stents.

In addition, Kim et al. [39] used oxidized hyaluronic acid and glycol chitosan as raw materials, and prepared bio-inks with self-healing ability in the presence of adipate dihydrazine. The imine bond formed by Schiff base reaction between oxidized hyaluronic acid and ethylene glycol chitosan, and the acylhydrazone bond formed between oxidized hyaluronic acid and adipic acid dihydrazine give the hydrogel bio-ink a good self-healing performance. This kind of bio-ink does not require an additional cross-linking process when preparing a 3D stent using extrusion 3D printing technology, and its self-healing ability, fluidity, and mechanical properties can be adjusted by adjusting the component ratio of the bio-ink. These characteristics can satisfy a wide range of applications in the field of tissue engineering (Figure 4.6).

Additive manufacturing is an emerging material processing method based on 3D CAD models, which can precisely control the microstructure and shape of objects. Compared with traditional methods, additive manufacturing can manufacture complex porous structures without using tools or molds. Zafeiris et al. [40] used additive

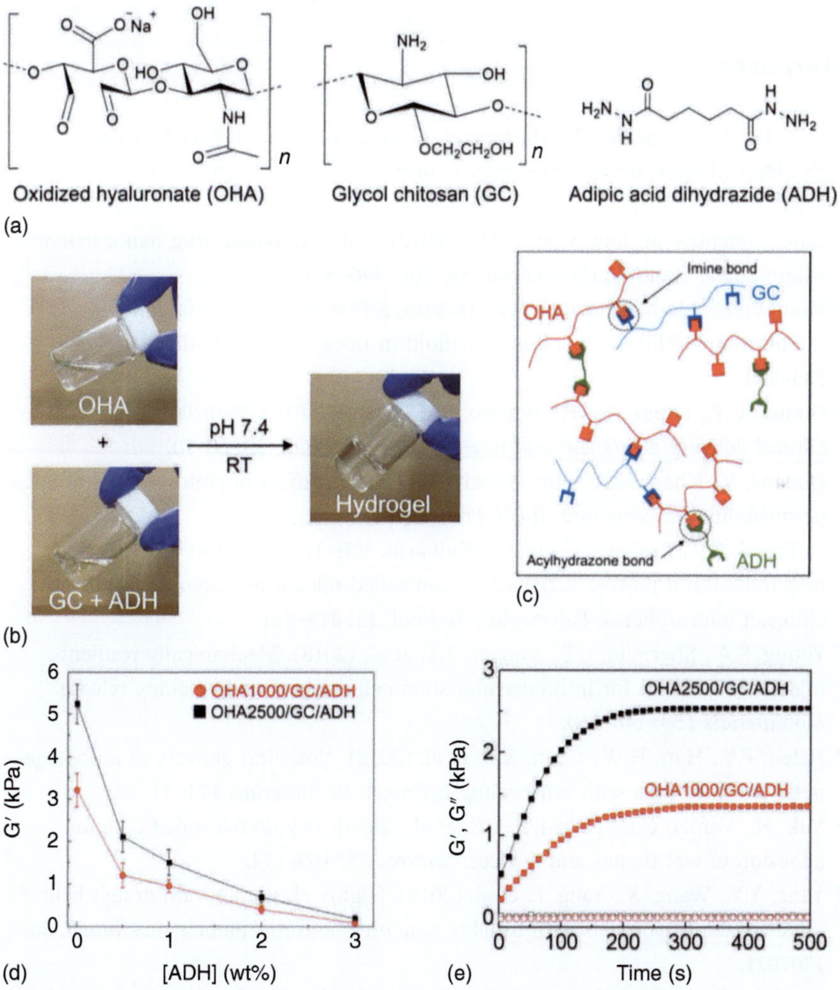

Figure 4.6 (a) Chemical structures of OHA, GC, and ADH. (b) Photographs of the polymer solution and the resulting hydrogel. (c) Schematic representation of the structure of the hydrogel. (d) Effect of ADH concentration on the storage modulus of the OHA/GC/ADH gel. (e) Kinetic process of gelation of the OHA/GC/ADH hydrogel. Source: Ref. [39]/with permission of American Chemical Society.

manufacturing technology to combine 3D printing with freeze-drying, and successfully prepared a tissue engineering scaffold with natural bone nanoporosity and interconnectivity. They used chitosan hydrogel as the matrix of the bio-ink and used L-arginine to adjust the nanocrystalline size of hydroxyapatite in the bio-ink component. To further reduce potential material toxicity, they use genipin as a cross-linking agent. After the printing process, the 3D scaffold is freeze-dried to remove the solvent in the package, thereby obtaining a porous interconnected network. Through microcomputer tomography and nanoindentation evaluation, it is confirmed that the porosity and modulus of the scaffold fall within the corresponding range of natural cancellous bone.

References

1 Coviello, T., Matricardi, P., Marianecci, C., and Alhaique, F. (2007). Polysaccharide hydrogels for modified release formulations. *J. Control. Release* 119: 5–24.

2 Garcia-Fuentes, M. and Alonso, M.J. (2012). Chitosan-based drug nanocarriers: where do we stand? *J. Control. Release* 161: 496–504.

3 VandeVord, P.J., Matthew, H.W.T., DeSilva, S.P. et al. (2002). Evaluation of the biocompatibility of a chitosan scaffold in mice. *J. Biomed. Mater. Res.* 59: 585–590.

4 Gomes, C.P., Lopes, C.D.F., Moreno, P.M.D. et al. (2014). Translating chitosan to clinical delivery of nucleic acid-based drugs. *MRS Bull.* 39: 60–70.

5 Dodane, V., Khan, M.A., and Merwin, J.R. (1999). Effect of chitosan on epithelial permeability and structure. *Int. J. Pharm.* 182: 21–32.

6 Kulkarni, P.V., Keshavayya, J., and Kulkarni, V.H. (2007). Effect of method of preparation and process variables on controlled release of insoluble drug from chitosan microspheres. *Polym. Adv. Technol.* 18: 814–821.

7 Young, S.A., Sherman, S.E., Cooper, T.T. et al. (2018). Mechanically resilient injectable scaffolds for intramuscular stem cell delivery and cytokine release. *Biomaterials* 159: 146–160.

8 Hsieh, F.Y., Han, H.W., Chen, X.R. et al. (2018). Non-viral delivery of an optogenetic tool into cells with self-healing hydrogel. *Biomaterials* 174: 31–40.

9 Yuk, H., Varela, C.E., Nabzdyk, C.S. et al. (2019). Dry double-sided tape for adhesion of wet tissues and devices. *Nature* 575: 169–174.

10 Yang, Y.Y., Wang, X., Yang, F. et al. (2018). Highly elastic and ultratough hybrid ionic-covalent hydrogels with tunable structures and mechanics. *Adv. Mater.* 30: 1707071.

11 Xu, L.J., Wang, C., Cui, Y. et al. (2019). Conjoined-network rendered stiff and tough hydrogels from biogenic molecules. *Sci. Adv.* 5: eaau3442.

12 Chedly, J., Soares, S., Montembault, A. et al. (2017). Physical chitosan microhydrogels as scaffolds for spinal cord injury restoration and axon regeneration. *Biomaterials* 138: 91–107.

13 Zhao, X., Wu, H., Guo, B.L. et al. (2017). Antibacterial anti-oxidant electroactive injectable hydrogel as self-healing wound dressing with hemostasis and adhesiveness for cutaneous wound healing. *Biomaterials* 122: 34–47.
14 Sheng, L.L., Zhang, Z.W.B., Zhang, Y. et al. (2021). A novel "hot spring"-mimetic hydrogel with excellent angiogenic properties for chronic wound healing. *Biomaterials* 264: 120414.
15 Wang, X.C., Ma, B., Xue, J.M. et al. (2019). Defective black nano-titania thermogels for cutaneous tumor-induced therapy and healing. *Nano Lett.* 19: 2138–2147.
16 Li, Y., Liu, X., Li, B. et al. (2020). Near-Infrared Light Triggered Phototherapy and Immunotherapy for Elimination of Methicillin-Resistant Staphylococcus aureus Biofilm Infection on Bone Implant. *ACS Nano* 14: 8157–8170.
17 Gao, G., Jiang, Y.W., Jia, H.R., and Wu, F.G. (2019). Near-infrared light-controllable on-demand antibiotics release using thermo-sensitive hydrogel-based drug reservoir for combating bacterial infection. *Biomaterials* 188: 83–95.
18 Mihic, A., Cui, Z., Wu, J. et al. (2015). A conductive polymer hydrogel supports cell electrical signaling and improves cardiac function after implantation into myocardial infarct. *Circulation* 132: 772–784.
19 He, S., Wu, J., Li, S.H. et al. (2020). The conductive function of biopolymer corrects myocardial scar conduction blockage and resynchronizes contraction to prevent heart failure. *Biomaterials* 258: 120285.
20 Man, Z.T., Hu, X.Q., Liu, Z.L. et al. (2016). Transplantation of allogenic chondrocytes with chitosan hydrogel-demineralized bone matrix hybrid scaffold to repair rabbit cartilage injury. *Biomaterials* 108: 157–167.
21 Lee, C.S., Hwang, H.S., Kim, S. et al. (2020). Inspired by nature: facile design of nanoclay-organic hydrogel bone sealant with multifunctional properties for robust bone regeneration. *Adv. Funct. Mater.* 30: 2003717.
22 Cui, Z.K., Kim, S., Baljon, J.J. et al. (2019). Microporous methacrylated glycol chitosan-montmorillonite nanocomposite hydrogel for bone tissue engineering. *Nat. Commun.* 10: 66.
23 Pan, W.Z., Dai, C.B., Li, Y. et al. (2020). PRP-chitosan thermoresponsive hydrogel combined with black phosphorus nanosheets as injectable biomaterial for biotherapy and phototherapy treatment of rheumatoid arthritis. *Biomaterials* 239: 119851.
24 Saudi, S., Bhattarai, S.R., Adhikari, U. et al. (2020). Nanonet-nano fiber electrospun mesh of PCL-chitosan for controlled and extended release of diclofenac sodium. *Nanoscale* 12: 23556–23569.
25 Shen, Y.B., Tu, T., Yi, B.C. et al. (2019). Electrospun acid-neutralizing fibers for the amelioration of inflammatory response. *Acta Biomater.* 97: 200–215.
26 Zhang, C., Yuan, H.H., Liu, H.H. et al. (2015). Well-aligned chitosan-based ultrafine fibers committed teno-lineage differentiation of human induced pluripotent stem cells for Achilles tendon regeneration. *Biomaterials* 53: 716–730.
27 Chen, J.W., Zhan, Y.F., Wang, Y.B. et al. (2019). Chitosan/silk fibroin modified nanofibrous patches with mesenchymal stem cells prevent heart remodeling post-myocardial infarction in rats (vol 80, pg 154, 2018). *Acta Biomater.* 89: 425–426.

28 Shalumon, K.T., Lai, G.J., Chen, C.H., and Chen, J.P. (2015). Modulation of bone-specific tissue regeneration by incorporating bone morphogenetic protein and controlling the shell thickness of silk fibroin/chitosan/nanohydroxyapatite core–shell nanofibrous membranes. *ACS Appl. Mater. Interfaces* 7: 21170–21181.

29 Huang, R., Li, W.Z., Lv, X.X. et al. (2015). Biomimetic LBL structured nanofibrous matrices assembled by chitosan/collagen for promoting wound healing. *Biomaterials* 53: 58–75.

30 Bazmandeh, A.Z., Mirzaei, E., Fadaie, M. et al. (2020). Dual spinneret electrospun nanofibrous/gel structure of chitosan-gelatin/chitosan-hyaluronic acid as a wound dressing: in vitro and in vivo studies. *Int. J. Biol. Macromol.* 162: 359–373.

31 Keirouz, A., Radacsi, N., Ren, Q. et al. (2020). Nylon-6/chitosan core/shell antimicrobial nanofibers for the prevention of mesh-associated surgical site infection. *J. Nanobiotechnol.* 18: 51.

32 Jing, X., Mi, H.Y., Wang, X.C. et al. (2015). Shish-Kebab-structured poly(ε-caprolactone) nanofibers hierarchically decorated with chitosan poly(ε-caprolactone) copolymers for bone tissue engineering. *ACS Appl. Mater. Interfaces* 7: 6955–6965.

33 Xu, T., Yang, H.Y., Yang, D.Z., and Yu, Z.Z. (2017). Polylactic acid nanofiber scaffold decorated with chitosan island like topography for bone tissue engineering. *ACS Appl. Mater. Interfaces* 9: 21094–21104.

34 Elviri, L., Foresti, R., Bergonzi, C. et al. (2017). Highly defined 3D printed chitosan scaffolds featuring improved cell growth. *Biomed. Mater.* 12: 045009.

35 Intini, C., Elviri, L., Cabral, J. et al. (2018). 3D-printed chitosan-based scaffolds: an in vitro study of human skin cell growth and an in-vivo wound healing evaluation in experimental diabetes in rats. *Carbohydr. Polym.* 199: 593–602.

36 Bozuyuk, U., Yasa, O., Yasa, I.C. et al. (2018). Light-triggered drug release from 3D-printed magnetic chitosan microswimmers. *ACS Nano* 12: 9617–9625.

37 Sayyar, S., Gambhir, S., Chung, J. et al. (2017). 3D printable conducting hydrogels containing chemically converted graphene. *Nanoscale* 9: 2038–2050.

38 Liu, Y., Wong, C.W., Chang, S.W., and Hsu, S.H. (2021). An injectable, self-healing phenol-functionalized chitosan hydrogel with fast gelling property and visible light-crosslinking capability for 3D printing. *Acta Biomater.* 122: 211–219.

39 Kim, S.W., Kim, D.Y., Roh, H.H. et al. (2019). Three-dimensional bioprinting of cell-laden constructs using polysaccharide-based self-healing hydrogels. *Biomacromolecules* 20: 1860–1866.

40 Zafeiris, K., Brasinika, D., Karatza, A. et al. (2021). Additive manufacturing of hydroxyapatite-chitosan-genipin composite scaffolds for bone tissue engineering applications. *Mater. Sci. Eng. C* 119: 111639.

5

Sources, Structures, and Properties of Other Polysaccharides

In addition to the natural polysaccharide polymers described earlier, there are other types of polysaccharides (such as starch and dextran) that are widely used in the field of biomedicine. In the next section, we will review the basic properties and latest research progress of these polysaccharides.

Starch is a high-molecular carbohydrate, a natural polysaccharide polymerized by glucose molecules, which is widely found in plant roots and crops such as corn, potatoes, and rice. The basic unit of starch is α-D-glucopyranose, which is divided into amylose and amylopectin according to its structure. Amylose is composed of D-hexacyclic glucose connected by α-(1-4)-glucosidic bond, the branching position of amylopectin is α-(1-6)-glucosidic bond, and the rest is α-(1-4) glycosidic bond [1–3]. Amylose contains hundreds of glucose units and amylopectin contains thousands of glucose units. For different sources of starch, its amylose content is also different, about 20–26%, and its particle shape is also different [4]. The content of amylose directly affects the gel ability of starch. The higher the content of amylose, the stronger the water-holding capacity. The starch modified by chemical (etherification, esterification, oxidation) and physical (gelatinization) processes around the three hydroxyl groups of starch has a wide range of applications in the biomedical field [5–7]. For example, hydroxyethyl starch solution plays an important role in expanding patients' blood volume. With the continuous development of hydrogel and electrospinning technology, the application of starch in biomedical field is also being further expanded.

Dextran is a monomer α-D-glucose polymer, which is a hydrophilic neutral polysaccharide. Dextran was first isolated as a microbial product by Louis Pasteur in 1861. At present, most of the strains producing glucan are *Candida albicans*. The main chain of glucan is connected by α-(1-6) glucosidic bond, and its branch chain is connected to the main chain by α-(1-4)-glucosidic bond or α-(1-3) glycosidic bond [8]. The source of dextran affects its molecular weight and branching degree. Dextran is well known to the public and it is still widely used in clinics as an excellent plasma substitute and was added to the list of basic plasma substitutes by the World Health Organization (WHO) in 2015 [9]. However, the application of dextran in biomedical field is much more than that. The sequence of dextran can be controlled by biological or chemical means, and the performance of dextran can be

Natural Polymers for Biomedical Applications, First Edition. Wenguo Cui and Lei Xiang.

further improved by modifying hydroxyl functional groups. Its application in drug delivery, tissue engineering, and regenerative medicine has been further expanded.

5.1 Other Polysaccharides-Based Hydrogel for Biomedical Application

5.1.1 Drug Delivery

Hydrogel is a three-dimensional cross-linked water-soluble polymer network, which can well coat different drugs and control their slow release, thereby providing higher local drug concentration in surrounding tissues for better treatment effect. However, it is difficult to achieve the controllable release of drugs for different microenvironments in vivo by simply using the coating effect of the hydrogel network. Based on the microenvironmental characteristics of human digestive tract rich in α-amylase, Davoodi-Monfared et al. [10] prepared an α-amylase-responsive hydrogel drug delivery system. Starch is a natural polysaccharide and does not have any functional groups suitable for producing stable starch-based hydrogels. Their starch is modified into carboxymethyl starch and cross-linked with $FeCl_3/FeCl_2$ solution. The synthesis was optimized under different conditions to prepare a hydrogel with suitable particle size. In vitro experiments have shown that α-amylase can significantly accelerate the degradation of starch-based hydrogels to promote the release of doxorubicin coated in the hydrogels. Compared with free drug molecules, the amylase-responsive hydrogel drug delivery system significantly enhances the anti-cancer performance in a specific site environment.

Injectable hydrogels are widely used in the field of minimally invasive topical drug delivery, but the dry closed tissue will adversely affect the flow properties of the hydrogel. Unterman et al. [11] added silicate nanosheets to the dextran-polyamide hydrogel, which significantly improved the intermolecular fluidity. In addition, they found that the aspect ratio of the nanosheets will significantly affect the physical properties of the hydrogel. For example, the nanosheets with a low aspect ratio will give the hydrogel stronger rheological properties, and the nanosheets with a high aspect ratio will enhance the water. Realizing the visualization of drug delivery is of great significance for understanding the metabolic process of drugs and understanding its mechanism of action. Currently commonly used chemically synthesized fluorescent dyes have certain cytotoxicity (Figure 5.1).

Liu et al. [12] prepared a hydrazone-cross-linked sericin/dextran injectable hydrogel using natural photoprotein sericin and dextran in silk as raw materials. This hydrogel has photoluminescence properties while achieving controlled drug release and can monitor the degradation of the hydrogel in vitro to calculate the release level of the drug. The drug loading of the hydrogel delivery system is closely related to the solubility of the drug in the aqueous solution, and it is also a challenging task to experiment with the sustained release of drugs under the premise of high drug loading. Dinh et al. [13] used microfluidic technology to separate PEGDA and dextran in pregel droplets to prepare drug-loaded microspheres with a core–shell structure. The drug is dispersed in the liquid core at a concentration higher than the saturation

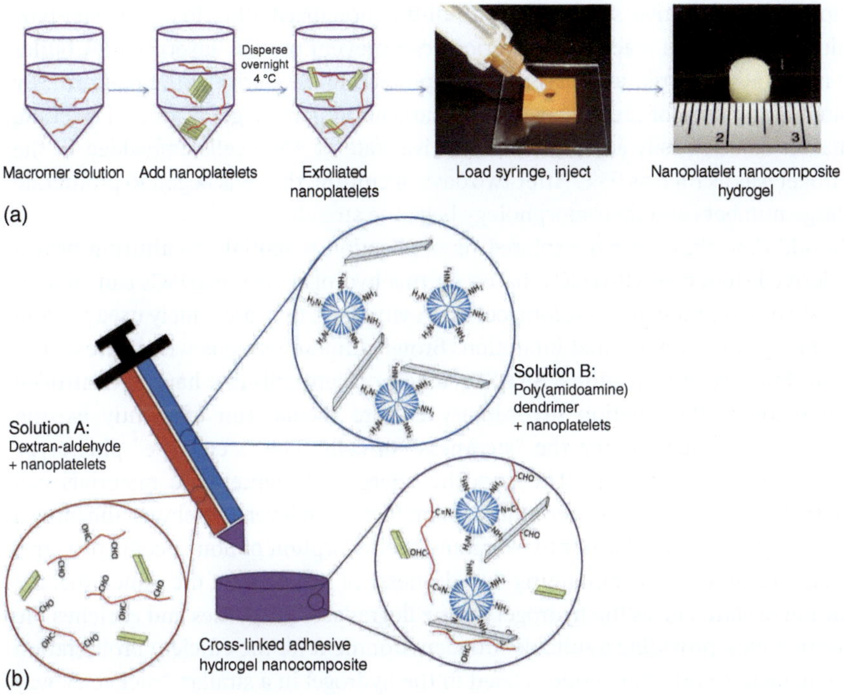

Figure 5.1 (a) Direct mixing of nanoparticle catalysts into large-molecular-weight water-soluble polymers, which are polymerized and gelatinized to form hydrogels containing nanofillers. (b) Mixing of dextran aldehydes and PAMAM polymer amine solutions, which are gelatinized by oxime bonding reactions, each can contain nanofillers.
Source: Ref. [11]/American Chemical Society.

solubility, and the drug is continuously administered by adjusting the polymer shell. This study can achieve a sustained drug release effect of up to 30 days on the basis of 80% drug loading. The purpose of controlled drug release is to prolong the action time of the drug on the one hand and to better control the blood drug concentration on the other hand. The plasma concentration of aminoglycosides directly affects their pharmacological effects, and high concentrations of aminoglycosides can cause irreversible ototoxicity. Hu Jingjing et al. [14] reported an on-demand administration method for aminoglycoside drugs. They used aminoglycosides as cross-linking agents to prepare a new type of aminoglycoside hydrogel. During the gel formation process, the modulus, degradation rate, and release kinetics of the hydrogel can be precisely adjusted by adjusting the amount of aminoglycoside drugs.

5.1.2 Cell and Organoid Culture

The three-dimensional cell embedding in hydrogel provides an excellent method for studying the biochemical clues that guide cell fate and extracellular behavior. The first step is to establish a suitable pure platform to simulate and simplify the natural extracellular matrix microenvironment. Dong et al. [15] prepared a "clickable" hydrogel based on zwitterionic starch using acylation-modified sulfobetaine-derivative starch and dithiol-functionalized PEG in a Michael-type "thiol-ene" addition reaction. Zwitterions can prevent protein adsorption by building a hydration barrier and play a role in protecting the purity of the platform. By changing the ratio of precursors, its mechanical properties, gel time, and swelling behavior can be easily adjusted. The survival rate of A549 cells embedded in the hydrogel was as high as 93%. After two days of culture, the cells began to proliferate in large numbers and their morphology began to stretch.

In addition, they further explored its application potential in culturing brown fat-derived stem cells (BADSCs) based on this hydrogel [16]. BADSCs can spontaneously differentiate into cardiomyocytes in vitro, and they are widely used in stem cell therapy after myocardial infarction. Brown adipose tissue is widely present in infants but very little in adults. BADSCs, a type of stem cells that have uncontrolled spontaneous differentiation in ordinary culture media, can efficiently expand BADSCs while maintaining the "stemness" of cells. The "stem-type" phenotype has become a top priority. The opposite charges of zwitterionic materials can electrostatically induce the surrounding high hydration layer, thereby establishing a high-energy hydration barrier to overcome the adsorption of non-specific proteins, which is conducive to maintaining the "dryness" of BADSCs. At the same time, the addition of starch gives the hydrogel matrix degradable properties and enriches the bioactive sites, providing a suitable microenvironment for the efficient proliferation of stem cells. CGRGDS peptide is fixed to the hydrogel in a similar "clickable" way to promote cell proliferation and adhesion behavior. On the basis of promoting the rapid proliferation of stem cells encapsulated in the hydrogel, its expanded cells also maintain efficient spontaneous myocardial differentiation.

Currently, most of the hydrogels used for cell embedding are based on nondegradable polyethylene glycol. Liu and Chan-Park [17] are committed to developing a cell culture system based on degradable natural polymers. They found that highly hydrophilic glucans are more similar to glycosaminoglycans, can better mimic the natural extracellular matrix, and can chemically modify functional groups to meet different cell culture needs. They recently reported the development of a composite hydrogel cell culture system of methacrylic acid and lysine-functionalized dextran and methacrylamide-modified gelatin. By adjusting the degree of functionalization and concentration of dextran and gelatin, the physical properties of the hydrogel can be further controlled.

5.1.3 Tissue Regeneration

The survival and functionalization of reconstructed tissues depend on whether they can generate vascular structures simultaneously, and the aggregation and growth of adipose stem cells (ASCs) are beneficial to promote angiogenesis. Xie et al. [18] proposed a porous hydrogel based on collagen-oxidized starch complex to promote the growth of adipose stem cells and stimulate angiogenesis. ASCs tend to attach to the hydrogel, spontaneously grow into spheres over time, and effectively promote the expression of vascular endothelial-cell-related growth factors. In addition, by adjusting the degree of oxidation of starch, the gelation time and pore size of the hydrogel can be controlled to meet the needs of different tissue engineering. Shock and multiple organ failure caused by bleeding are still a major direct cause of death.

At present, starch-based hemostatic powder is often used in the treatment of various types of clinical acute and chronic bleeding, but this hemostatic powder is easy to fall off the surface of the tissue in the body fluid environment, which limits the hemostatic effect. Cui et al. [19] were inspired by muscle adhesion protein and prepared a new injectable tissue adhesion hydrogel composed of starch, succinic anhydride, and dopamine. This kind of hydrogel can be tightly combined with the surface of the tissue to physically block the bleeding site to avoid serious consequences caused by heavy bleeding. Liu et al. [20] combined aldehyde hydroxyethyl starch with aldehyde group and amino carboxymethyl chitosan with amino group to prepare a two-component hemostatic hydrogel through Schiff base reaction. The mechanical properties of the hydrogel can be controlled by adjusting the content of amino and aldehyde groups in the two components. This kind of hydrogel can not only physically block the bleeding port, but the negatively charged amino groups in the hydrogel can also specifically bind to the positively charged platelets in the blood to activate the blood coagulation process in the body. Also based on the Schiff base reaction, Li et al. [21] used oxidized hydroxyethyl starch and modified carboxymethyl chitosan as raw materials to prepare a good compliance inject the hydrogel and apply it to wound healing. By changing the ratio of the two components, they found that the oxidized hydroxyethyl starch and modified carboxymethyl chitosan hydrogel with a volume ratio of 5:5 has a suitable gelation time, significant water retention capacity, and good self-healing ability. In the healing experiment of full-thickness skin defects in rats, the oxidized hydroxyethyl starch and modified carboxymethyl chitosan hydrogel treatment group with good moisturizing ability increased the wound closure rate, increased the formation of granulation tissue, and accelerated the speed of epithelialization (Figure 5.2).

The role of geometric constraints and cell–matrix mechanical interactions in the formation of capillaries has not been elucidated. Sun et al. [22] achieved geometric control of the endothelial network topology by using micro-fences and wells to create physical constraints. By adjusting the concentration of PEG–dextran hydrogel to change the hardness of the matrix, the effect of mechanical interaction on capillary formation was explored. They found that capillary-like structures are only formed on compliant substrates (E = 200–1000 Pa), and there are almost no capillary-like structures formed on hard substrates, confirming that capillaries can be regulated by the mechanical interaction of cell–matrix formation process of blood vessels. In addition, Wei and Gerecht [23] prepared an injectable hydrogel with self-healing ability based on the dynamic imine cross-linking between gelatin and oxidized dextran. This kind of hydrogel can protect the progenitor cells from the shear force of the injection and allows the cells to be placed in a controlled space/time. In addition, a large number of biologically active sites in the hydrogel can support the formation of functionalized vascular structures by endothelial progenitor cells.

Adding various metal nanoparticles (such as Zn, Cu, Ag) to the hydrogel system is a common method for preparing hydrogels with antibacterial properties. The

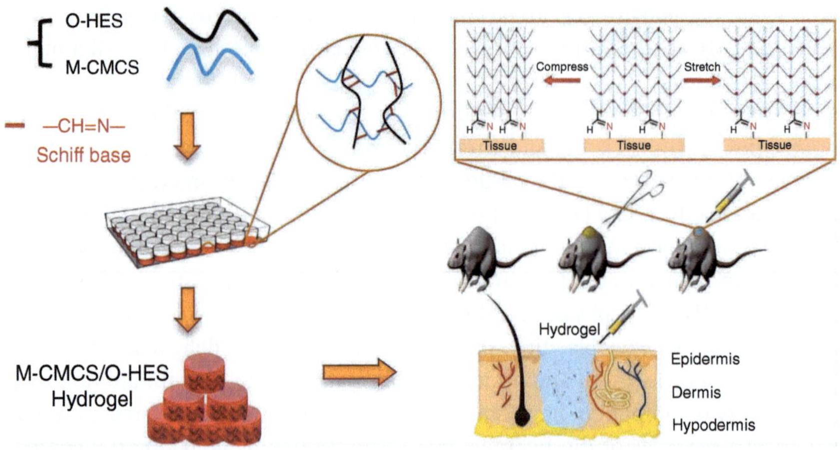

Figure 5.2 An in situ molded hydrogel material with good biocompatibility, biodegradability, and self-recovery properties for the treatment of total skin injury was obtained by optimizing the ratio of oxidized hydroxyethyl starch and modified carboxymethyl chitosan using Schiff base reaction. Source: Ref. [21]/American Chemical Society.

addition of metal nanoparticles often has a certain impact on the swelling properties of the hydrogel. For this reason, Namazi et al. [24] thoroughly discussed the influence of nano-zinc oxide on the swelling behavior of oxidized starch hydrogels. The results show that the swelling behavior of the hydrogel is related to the acid–base environment. When the pH is 7, the swelling degree of the hydrogel is the highest due to the formation of the carboxylic anion; in various salt solutions, the swelling rate of the hydrogel decreases with the salt solution concentration and cation valence increase; in addition, the swelling capacity of the hydrogel is positively correlated with the abundance of zinc oxide nanoparticles. At the same time, due to the addition of zinc oxide nanoparticles, the prepared hydrogel exhibits good antibacterial properties against *Staphylococcus aureus* and *Escherichia coli*. Villanueva et al. [25] used copper salts and hydrazine to prepare nanoparticles in a starch medium in order to improve the stability of metal nanoparticles in hydrogels and then coated them with silica. Starch caps are formed around to further enhance the stability of copper nanoparticles. Acute skin toxicity tests show that the stability of copper nanoparticles has been significantly improved, and they are only slightly irritating to the skin. Starch-copper nanoparticle hydrogels show long-term antibacterial activity against Gram-negative bacteria and Gram-positive bacteria. Cationic small-molecule fungicides have been developed as promising anti-biofilm preparations due to their adjustable chemical structure and ability to destroy established biofilms. In order to improve its sustained-release process at the target site, Hoque et al. [26] reported a new type of antibacterial system with dextran methacrylate hydrogel loaded with a cationic fungicide.

In addition, for chronic infection wounds, Wei et al. [27] mixed oxidized dextran, antimicrobial peptide-modified hyaluronic acid, and PRP under physiological conditions based on the formation of Schiff base bonds. The oxidized dextran/antimicrobial peptide-modified hyaluronic acid/PRP hydrogel was prepared. While achieving significant inhibition of *E. coli*, *S. aureus*, and *Pseudomonas aeruginosa*, platelet-rich plasma can significantly promote collagen deposition, vascular deposition, and wound repair.

5.2 Other Polysaccharides-Based Electrospinning for Biomedical Application

5.2.1 Drug Delivery

Electrospun starch fiber is a biological material with broad application prospects in the fields of drug delivery and tissue engineering. There are two forms of natural starch, amylose and amylopectin, and the difference in the ratio between the two will give electrospun starch fibers different characteristics. Vasilyev et al. [28] used different amylose and amylopectin blends to characterize the behavior of starch in formic acid, and explored the effect of starch blends with different component ratios in electrospinning. They found that even pure amylopectin solution can be electrospun, and the higher the amylopectin content, the more brittle the electrospun fiber. The results of the study revealed that fibers with higher amylose content are suitable for tissue engineering, while fibers with higher amylopectin content have more potential for drug-controlled release. Growth factors are often used in the treatment of various diseases, and the feasibility of using electrospun membranes as a growth-factor-controlled release carrier has also been extensively explored. However, growth factors very easily lose their activity during the electrospinning process, which limits their application in biomedicine. Liu et al. [29] first used dextran glass nanoparticles as the release carrier of basic fibroblast growth factor to ensure the biological activity of the growth factor, and then co-spun it with poly-L-lactic acid to prepare electrospun membrane. The results show that the dual effects of spinning membrane encapsulation and slow release of growth factors can effectively prevent tendon adhesion and promote tendon healing.

5.2.2 Tissue Regeneration

The formation of nanofibers by electrospinning technology can provide a moist environment and avoid dehydration of the wound surface, so it has received great attention. In addition, the nanometer size of nanofibers naturally makes it have a high specific surface-area-to-volume ratio, which can significantly enhance cell adhesion, proliferation, migration, and differentiation functions, and its high porosity facilitates the exchange and transportation of nutrients. The content of starch will affect the structural stability of the overall fiber. In order to better solve this problem, Vasilyev et al. [28] used coaxial electrospinning technology to use polyurethane as the structural core of nanofibers, and a mixture of starch and hyaluronic acid as a bioactive shell to promote wound healing. Whether it is an in vitro experiment based on mouse fibroblasts or a wound healing experiment, this putamen structure starch

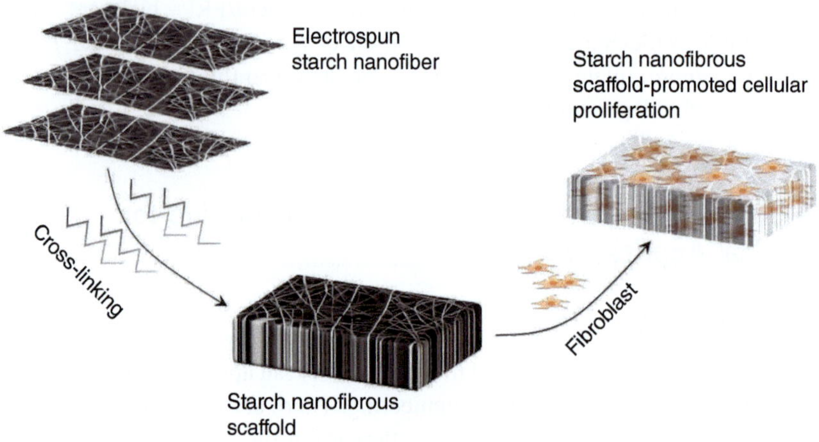

Figure 5.3 Preparation process of starch-based nanofibers. Source: Ref. [31]/Elsevier.

electrospun fiber shows great potential to promote wound healing. In addition, the uncross-linked electrospun starch fiber is a highly hydrophobic material, and the fiber will swell when immersed in water and cause the destruction of the overall structure. Using glutaraldehyde as a cross-linking agent to cross-link the fibers can significantly improve the stability of electrospun fibers in water, thereby further improving the performance of electrospun starch fibers in promoting wound healing [30, 31] (Figure 5.3).

Dextran can promote angiogenesis on chronic wounds and induce platelet adhesion. It can be blended with suitable polymers such as polyurethane to prepare composite nanofiber membranes by electrospinning. Unnithan et al. [32] loaded anti-inflammatory estrogen on a dextran electrospun membrane. Under the dual action of dextran and estrogen, the inflammatory response of the wound was effectively inhibited and the wound healing speed was also improved. Sagitha et al. [33] further studied the properties of dextran–polyurethane fiber membranes. They found that after adding dextran to polyurethane, its hydrophilicity, moisture permeability, water absorption, and biodegradability were all improved. They loaded curcumin on the dextran-polyurethane electrospun membrane, which can control the release of drugs by adjusting the acid–base environment.

Starch can mimic natural ECM, and starch is an indispensable energy storage substance for the human body. Starch can be metabolized into glucose in the body to provide energy for the body and at the same time provide a source of carbohydrates for tissue repair. The large number of hydroxyl groups contained in starch provides many functional groups that can be used for artificial chemical modification. However, the mechanical properties of unmodified starch materials are difficult to meet the needs of bone tissue repair engineering, so how to further improve the mechanical properties of starch materials has become a key research object. Wu et al. [34] further enhanced the mechanical properties of electrospun starch fibers by adding graphene oxide as a nano-strengthening agent and using formic acid as a starch solvent and esterification agent. At the same time, the addition of graphene oxide further improves the electrospun nuclear fiber morphology of the starch solution and can induce the mineralization of hydroxyapatite to accelerate the bone tissue repair process.

In addition, blending with other polymers is also a way to improve the performance of electrospun nanofibers in the field of bone tissue repair. Gutiérrez-Sánchez et al. [35] blended starch and polylactic acid to prepare electrospun fiber mats. The arginine–glycine–aspartic acid polypeptide (RGD) is added to the surface of the fiber mat by physical adsorption to improve cell adhesion. Experimental results show that fiber mats with 5.0% starch content can create a good external environment for osteoblast proliferation. In tissue engineering, how to synchronize the rate of degradation with the rate of natural tissue formation is a major challenge in material design. Movahedi et al. [36] proposed the idea of using electron beams to induce degradation of starch/polylactic acid/β-tricalcium phosphate electrospun scaffolds. Because β-tricalcium phosphate has a rough surface morphology, it can promote the attachment of bone cells and is used as the osteogenic activator of the composite scaffold. They found that the mechanical strength of the fiber's water contact angle and nucleus will show a downward trend with the increase of the irradiation dose, which may be caused by the fracture of the fiber structure caused by the electron beam irradiation with a high irradiation dose. This discovery provides a new idea for preparing bone repair scaffolds with controllable degradation properties.

5.3 Other Polysaccharides 3D Printing for Biomedical Application

5.3.1 Drug Delivery

Due to their degradability, low immunogenicity, and good biocompatibility, 3D printing biomaterials using starch as raw materials are widely used in various fields of biomedicine. 3D printed stents using porous hydrogel as bio-inks show unlimited potential in controlling the release of drugs. Bom et al. [37] used extrusion 3D printing technology to prepare alginate drug delivery patches and explored the effect of pregelatinized starch on the performance of this drug delivery system. They found that the addition of pregelatinized starch can increase the number and diameter of the holes in the structure, and by changing the amount of starch added, the mechanical properties, printing accuracy, and drug release phase of the hydrogel patch can be further controlled.

5.3.2 Tissue Regeneration

It has been proved that the stability of 3D printed structures can be improved through cross-linking, but commonly used chemical cross-linking methods may bring potential toxicity to biomaterials, so photocross-linking methods are favored by more and more researchers. Noè et al. [38] used methacrylic anhydride to modify starch and successfully prepared a hydrogel bio-ink with photocross-linking properties. By changing the concentration of the starch solution (for example, from 10 to 15 wt%), the compression stiffness of the hydrogel can be effectively controlled (from 13 to 20 kPa). Methacrylated starch bio-ink can meet the hardness requirements of various tissues of the human body and further expand its application prospects.

Since the gel time of bio-ink is difficult to accurately control, problems such as ink clogging of nozzles or collapse of printed structures are often encountered during the 3D printing process. Based on this problem, Du et al. [39] successfully prepared 3D printing bio-inks with controllable gelation time based on the phase separation of gelatin and oxidized dextran. Positively and negatively charged gelatin can drive the phase separation process of polysaccharides through orderly self-aggregation, thereby inhibiting the nucleophilic addition reaction between aldehydes on the oxidized dextran chains and amines on the gelatin chains. By adjusting the pH value, the amino groups of the gelatin are protonated or deprotonated, thereby adjusting the interfacial cross-linking kinetics between the gelatin-rich phase and the dextran-rich phase, so as to achieve the effect of controlling the gelation time. Coupled with the heat-sensitizing properties of gelatin, this bio-ink can well meet the needs of extrusion 3D printing.

References

1 Liu, J., Wang, X.C., Pu, H.M. et al. (2017). Recent advances in endophytic exopolysaccharides: production, structural characterization, physiological role and biological activity. *Carbohydr. Polym.* 157: 1113–1124.

2 Wang, S.J., Li, C.L., Copeland, L. et al. (2015). Starch retrogradation: a comprehensive review. *Compr. Rev. Food Sci. Food Saf.* 14: 568–585.

3 Lu, D.R., Xiao, C.M., and Xu, S.J. (2009). Starch-based completely biodegradable polymer materials. *Express Polym. Lett.* 3: 366–375.

4 Singh, N., Singh, J., Kaur, L. et al. (2003). Morphological, thermal and rheological properties of starches from different botanical sources. *Food Chem.* 81: 219–231.

5 Wang, S.J. and Copeland, L. (2015). Effect of acid hydrolysis on starch structure and functionality: a review. *Crit. Rev. Food Sci. Nutr.* 55: 1079–1095.

6 Hong, Y., Liu, G.D., and Gu, Z.B. (2016). Recent advances of starch-based excipients used in extended-release tablets: a review. *Drug Deliv.* 23: 12–20.

7 Liu, G.D., Hong, Y., Gu, Z.B. et al. (2015). Preparation and characterization of pullulanase debranched starches and their properties for drug controlled-release. *RSC Adv.* 5: 97066–97075.

8 Takashima, Y., Fujita, K., Ardin, A.C. et al. (2015). Characterization of the dextran-binding domain in the glucan-binding protein C of *Streptococcus mutans*. *J. Appl. Microbiol.* 119: 1148–1157.

9 Zheng, F.F., Chen, H.D., Chen, Y.F. et al. (2018). Comparative analysis of ADR on China's National Essential Medicines List (2015 Edition) and WHO Model List of Essential Medicines (19th Edition). *Biomed. Res. Int.* 2018: 7862036.

10 Davoodi-Monfared, P., Akbari-Birgani, S., Mohammadi, S. et al. (2021). Synthesis, characterization, and in vitro evaluation of the starch-based α-amylase responsive hydrogels. *J. Cell. Physiol.* 236: 4066–4075.

11 Unterman, S., Charles, L.F., Strecker, S.E. et al. (2017). Hydrogel nanocomposites with independently tunable rheology and mechanics. *ACS Nano* 11: 2598–2610.

12 Liu, J., Qi, C., Tao, K.X. et al. (2016). Sericin/dextran injectable hydrogel as an optically trackable drug delivery system for malignant melanoma treatment. *ACS Appl. Mater. Interfaces* 8: 6411–6422.

13 Dinh, N.D., Kukumberg, M., Nguyen, A.T. et al. (2020). Functional reservoir microcapsules generated *via* microfluidic fabrication for long-term cardiovascular therapeutics. *Lab Chip* 20: 2756–2764.

14 Hu, J.J., Quan, Y.C., Lai, Y.P. et al. (2017). A smart aminoglycoside hydrogel with tunable gel degradation, on-demand drug release, and high antibacterial activity. *J. Control. Release* 247: 145–152.

15 Dong, D.Y., Li, J.J., Cui, M. et al. (2016). In situ "clickable" zwitterionic starch-based hydrogel for 3D cell encapsulation. *ACS Appl. Mater. Interfaces* 8: 4442–4455.

16 Dong, D., Hao, T., Wang, C. et al. (2018). Zwitterionic starch-based hydrogel for the expansion and "stemness" maintenance of brown adipose derived stem cells. *Biomaterials* 157: 149–160.

17 Liu, Y.X. and Chan-Park, M.B. (2010). A biomimetic hydrogel based on methacrylated dextran-graft-lysine and gelatin for 3D smooth muscle cell culture. *Biomaterials* 31: 1158–1170.

18 Xie, X.F., Li, X.Y., Lei, J.F. et al. (2020). Oxidized starch cross-linked porous collagen-based hydrogel for spontaneous agglomeration growth of adipose-derived stem cells. *Mater. Sci. Eng. C* 116: 111165.

19 Cui, R.H., Chen, F.P., Zhao, Y.J. et al. (2020). A novel injectable starch-based tissue adhesive for hemostasis. *J. Mater. Chem. B* 8: 8282–8293.

20 Liu, J., Li, J., Yu, F. et al. (2020). In situ forming hydrogel of natural polysaccharides through Schiff base reaction for soft tissue adhesive and hemostasis. *Int. J. Biol. Macromol.* 147: 653–666.

21 Li, J., Yu, F., Chen, G. et al. (2020). Moist-retaining, self-recoverable, bioadhesive, and transparent in situ forming hydrogels to accelerate wound healing. *ACS Appl. Mater. Interfaces* 12: 2023–2038.

22 Sun, J., Jamilpour, N., Wang, F.Y., and Wong, P.K. (2014). Geometric control of capillary architecture via cell-matrix mechanical interactions. *Biomaterials* 35: 3273–3280.

23 Wei, Z. and Gerecht, S. (2018). A self-healing hydrogel as an injectable instructive carrier for cellular morphogenesis. *Biomaterials* 185: 86–96.

24 Namazi, H., Hasani, M., and Yadollahi, M. (2019). Antibacterial oxidized starch/ZnO nanocomposite hydrogel: synthesis and evaluation of its swelling behaviours in various pHs and salt solutions. *Int. J. Biol. Macromol.* 126: 578–584.

25 Villanueva, M.E., Diez, A.M.D., González, J.A. et al. (2016). Antimicrobial activity of starch hydrogel incorporated with copper nanoparticles. *ACS Appl. Mater. Interfaces* 8: 16280–16288.

26 Hoque, J. and Haldar, J. (2017). Direct synthesis of dextran-based antibacterial hydrogels for extended release of biocides and eradication of topical biofilms. *ACS Appl. Mater. Interfaces* 9: 15975–15985.

27 Wei, S.K., Xu, P.C., Yao, Z.X. et al. (2021). A composite hydrogel with co-delivery of antimicrobial peptides and platelet-rich plasma to enhance healing of infected wounds in diabetes. *Acta Biomater.* 124: 205–218.

28 Vasilyev, G., Vilensky, R., and Zussman, E. (2019). The ternary system amylose-amylopectin-formic acid as precursor for electrospun fibers with tunable mechanical properties. *Carbohydr. Polym.* 214: 186–194.

29 Liu, S., Qin, M.J., Hu, C.M. et al. (2013). Tendon healing and anti-adhesion properties of electrospun fibrous membranes containing bFGF loaded nanoparticles. *Biomaterials* 34: 4690–4701.

30 Mistry, P., Chhabra, R., Muke, S. et al. (2021). Fabrication and characterization of starch-TPU based nanofibers for wound healing applications. *Mater. Sci. Eng. C* 119: 111316.

31 Waghmare, V.S., Wadke, P.R., Dyawanapelly, S. et al. (2018). Starch based nanofibrous scaffolds for wound healing applications. *Bioact. Mater.* 3: 255–266.

32 Unnithan, A.R., Sasikala, A.R.K., Murugesan, P. et al. (2015). Electrospun polyurethane-dextran nanofiber mats loaded with estradiol for post-menopausal wound dressing. *Int. J. Biol. Macromol.* 77: 1–8.

33 Sagitha, P., Reshmi, C.R., Sundaran, S.P. et al. (2019). In-vitro evaluation on drug release kinetics and antibacterial activity of dextran modified polyurethane fibrous membrane. *Int. J. Biol. Macromol.* 126: 717–730.

34 Wu, D., Samanta, A., Srivastava, R.K., and Hakkarainen, M. (2017). Starch-derived nanographene oxide paves the way for electrospinnable and bioactive starch scaffolds for bone tissue engineering. *Biomacromolecules* 18: 1582–1591.

35 Gutiérrez-Sánchez, M., Escobar-Barrios, V.A., Pozos-Guillén, A., and Escobar-García, D.M. (2019). RGD-functionalization of PLA/starch scaffolds obtained by electrospinning and evaluated *in vitro* for potential bone regeneration. *Mater. Sci. Eng. C* 96: 798–806.

36 Movahedi, M., Asefnejad, A., Rafienia, M., and Khorasani, M.T. (2020). Potential of novel electrospun core-shell structured polyurethane/starch (hyaluronic acid) nanofibers for skin tissue engineering: in vitro and in vivo evaluation. *Int. J. Biol. Macromol.* 146: 627–637.

37 Bom, S., Santos, C., Barros, R. et al. (2020). Effects of starch incorporation on the physicochemical properties and release kinetics of alginate-based 3D hydrogel patches for topical delivery. *Pharmaceutics* 12: 719.

38 Noè, C., Tonda-Turo, C., Chiappone, A. et al. (2020). Light processable starch hydrogels. *Polymers (Basel)* 12: 1359.

39 Du, Z.M., Li, N.F., Hua, Y.J. et al. (2017). Physiological pH-dependent gelation for 3D printing based on the phase separation of gelatin and oxidized dextran. *Chem. Commun.* 53: 13023–13026.

6

Summary

Polysaccharide is a kind of natural polymer which is rich in nature and is composed of monosaccharides connected by glycosidic bonds. Different units and branching structures of polysaccharides lead to many physical and chemical properties of polysaccharides. Polysaccharides are commonly found in algae (alginate), plants (starch, cellulose), microorganisms (glucan), and animals (chitosan, hyaluronic acid). Hyaluronic acid, alginate, and chitosan are typical pH-sensitive substances in polysaccharides, and their physicochemical properties will change under different pH conditions. Based on this, many pH-corresponding hydrogels and drug delivery systems have been developed. In addition, due to the excellent mechanical properties of polysaccharides such as cellulose and starch, they are also widely used in the preparation of hydrogels, 3D printing, and electrospinning. The nano-forms of all kinds of polysaccharides also have excellent performance in drug sustained release and targeted delivery.

Natural Polymers for Biomedical Applications, First Edition. Wenguo Cui and Lei Xiang.

Section III

Polypeptides for Biomedical Application

Natural Polymers for Biomedical Applications, First Edition. Wenguo Cui and Lei Xiang.

7

Sources, Structures, and Properties of Collagen

Collagen is a kind of protein family which exists widely in animal connective tissue and is an important part of extracellular matrix. Up to now, at least 30 kinds of collagen chain coding genes have been found, which can form more than 16 kinds of collagen molecules.

Collagen can be divided into two types according to its function. The first group is fibroblast collagen, which mainly includes type I, II, and III collagen, and the rest is non-fibroblast collagen, mainly including type IV and V collagen. Fibroblast collagen accounts for about 90% of the total collagen, and type I collagen, which is widely used in biomedical field, belongs to fibroblast collagen. The detailed classification of various collagens has been described elsewhere and will not be discussed again [1].

In the middle of the twentieth century, with the development of X-ray diffraction technology and the proposal of amino acid sequence model, the triple helix structure of collagen [2] was discovered for the first time, that is, three polypeptide chains formed a left-handed helix structure and then bind each other with hydrogen bonds to form a solid right-handed superhelix structure. The amino acids of collagen are arranged periodically $(\text{Gly-X-Y})_n$, in which X and Y positions are proline and hydroxyproline are the unique amino acids of collagen. The tight triple helix structure of collagen has very low immunogenicity and high structural stability, and can bind to cell receptors to regulate cell adhesion, proliferation, and migration [3].

In order to further improve the properties of collagen and expand its application in biomedical field, researchers often use physical or chemical cross-linking methods to further improve its mechanical and biological properties [4–6]. At present, collagen-derived biomaterials and drug delivery systems have been widely studied.

7.1 Collagen-Based Hydrogel for Biomedical Application

7.1.1 Drug Delivery

The design of drug delivery according to the changes of the microenvironment of the lesion site is the mainstream idea to optimize drug delivery recently. In the case of myocardial infarction, the expression of MMP-2/9 is increased after myocardial infarction. Based on this point, Fan et al. [7] fused glutathione transferase (GST) and

Natural Polymers for Biomedical Applications, First Edition. Wenguo Cui and Lei Xiang.

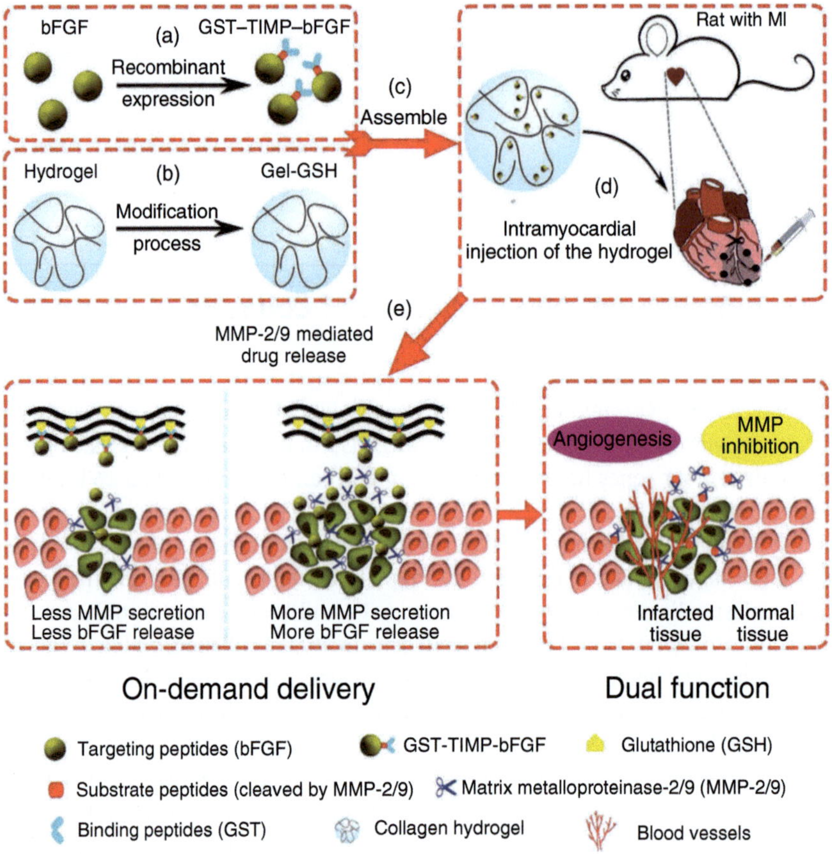

Figure 7.1 (a) Expression and purification of recombinant fusion protein GST–TIMP–bFGF. (b) Chemical cross-linking of GSH into hydrogel to prepare Gel-GSH. (c) Mixing of GST–TIMP–bFGF and Gel-GSH, which was bound by the GST–GSH interbonds. (d) Injection of the mixed hydrogel into the myocardium of post-infarcted rat myocardium. (e) Peptide was degraded by MMP at the wound site. (e) At the wound site, the peptide was degraded by MMP and released, realizing the dual functions of angiogenesis and MMP inhibition. Source: Ref. [7]/John Wiley & Sons.

matrix metalloproteinase-2 shock 9 (MMP-2/9) cleavage peptide PLGLAG (TIMP) with basic fibroblast growth factor to couple collagen amine group with glutathione (GSH) sulfhydryl group. GSH modified collagen hydrogel, and recombinant protein GST–TIMP–bFGF were prepared. The specific combination of GST and GSH significantly increased the drug loading of GST–TIMP–bFGF in collagen–GSH hydrogel. The TIMP peptide wrapped between GST and bFGF reacts to MMPs to release as needed during myocardial infarction. In addition, TIMP peptide is a competitive substrate of MMPs, which can inhibit the excessive degradation of cardiac matrix by MMPs after myocardial infarction. Hydrogels are also often used as carriers of cell therapy, but most studies have focused on how to maximize the activity of the cells carried, ignoring the problem of how to terminate the activity of the cells carried in an emergency (Figure 7.1).

Wong [8] achieved safer drug delivery by constructing a composite collagen–alginate saline gel with a Tet-on-inducible pre-Caspase8 mechanism as an oral-induced biosafety switch. Both in vitro and in vivo experiments showed that drug delivery was terminated by oral doxycycline without destroying the structure of the gel.

7.1.2 Cell and Organoid Culture

Collagen hydrogel is often used for cell 3D culture, and it can be combined with various growth factors to provide a suitable microenvironment for cell proliferation and differentiation. There are also many research and innovations that bionic the structure of specific human body parts, and regulate the behavior of cells from both physical and chemical aspects. Among them, the biomimetic cell culture collagen hydrogel designed for the structure of the intestinal crypt has received extensive attention. Hinman et al. [9] used primary human intestinal epithelial stem cells to form crypts on a 3D-shaped hydrogel scaffold to replicate the function and structural characteristics of crypts in vivo, and the ends of the hydrogel support the cross-over crypts. The formation of chemical concentrations of growth factors, inflammatory substances, and bacterial products better mimics the impact of the microenvironment in the body on cells.

7.1.3 Cell Behavior Regulation

Controlling the directional differentiation of stem cells is a hot research area of tissue regeneration. The emergence of collagen hydrogels has given more possibilities for the directional differentiation of stem cells. Bone marrow mesenchymal stem cells (BMSCs)-directed differentiation into cartilage is the main method of cartilage repair at present, but the directed differentiation tendency is poor and the differentiation efficiency is low. In order to overcome these two obstacles, Zheng et al. [10] constructed an injectable composite hydrogel with collagen hydrogel as a scaffold to accommodate bone marrow mesenchymal stem cells, cadmium selenide, and quantum dots. The introduction of CdSe quantum dots greatly enhances the hardness of collagen hydrogels through the mutual cross-linking of natural cross-linking agents, at the same time, triggers photodynamic excitation to generate ROS. Experimental results show that increased stiffness and increased production of reactive oxygen species can synergistically promote the proliferation of BMSCs, induce cartilage-specific gene expression, and increase the secretion of glycosaminoglycans (Figure 7.2).

In the field of nerve regeneration, the combination of mesenchymal stem cells and collagen hydrogel has been proven to have excellent activity in anti-inflammation and promoting neuron regeneration in the injured area [11]. This dual function is to rely on the self-assembly of MSCs in the spherical collagen hydrogel to effectively improve the internal contact between cells and the interaction between cells and the extracellular matrix. Strong cell communication promotes the secretion of endogenous nutritional factors and extracellular matrix components, which may help promote the differentiation of neural stem cells into neurons. In addition, collagen-MSC

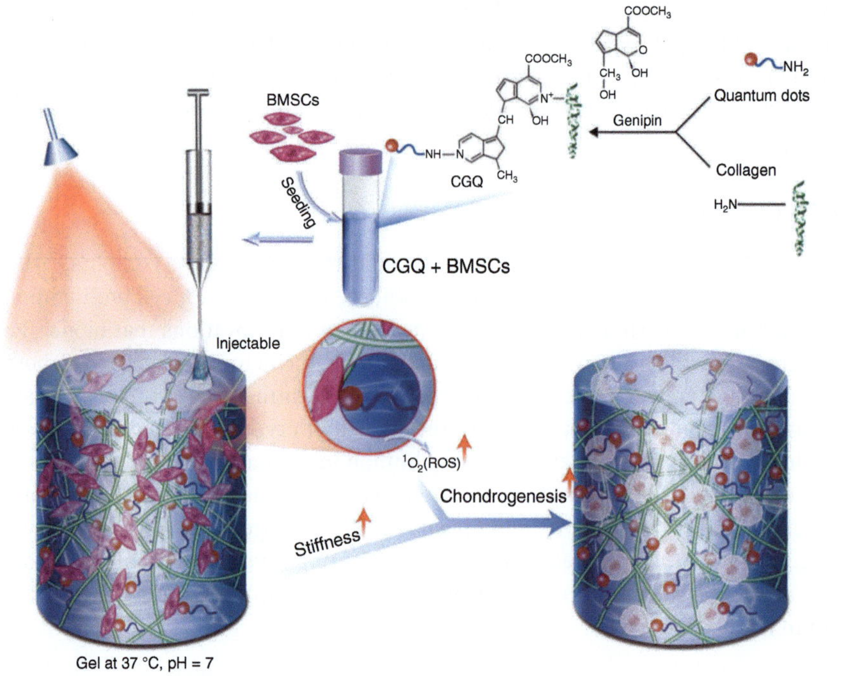

Figure 7.2 General schematic illustration of the fabrication process and implementation of collagen–genipin–quantum dot (CGQ) composite hydrogels. Source: Ref. [10]/John Wiley & Sons/CC BY 4.0.

sphere hydrogel creates an immune activation environment to inhibit LPS-induced inflammatory response by promoting the secretion of cytokines related to innate immune response and adaptive immune response.

7.1.4 Tissue Regeneration

Collagen is the main component of extracellular matrix and is arranged in different levels to form the main rigid component in connective tissue. Mredha et al. [12] used fish capsule collagen extracted from sturgeon, and, based on the concept of double network, successfully developed a new type of collagen-fiber-based flexible hydrogel. Double-network hydrogel consists of physically/chemically cross-linked anisotropic collagen fibers as the first network and neutral biocompatible poly(*N*,*N*-dimethylacrylamide) (PDMAAm) as the second network. Utilizing the excellent fiber-forming ability of collagen, the free injection method is adopted to form an anisotropic structure of the collagen fibril network with good retention in the double-network hydrogel. Double-network hydrogel increases the denaturation temperature of collagen. These double-network gels have anisotropic swelling behavior and exhibit excellent mechanical properties comparable to natural cartilage. In the field of ocular tissue repair, collagen can well mimic the corneal stroma and can reduce the incidence of complications related to the donor cornea, including infection and rejection. Chen et al. [13] published a study on bio-orthogonal cross-linked hyaluronic-acid–collagen hydrogel, which can fill corneal defects in situ without additional sutures. This hydrogel in vitro and in vivo shows good cell compatibility and support for epithelialization.

7.2 Collagen-Based Electrospinning for Biomedical Application

7.2.1 Cell and Organoid Culture

How to maintain the characteristics of cells in vitro is a major difficulty in the field of biomedicine. Taking primary hepatocytes as an example, their cytochrome P450 activity will be progressively lost when cultured in vitro. For this reason, Brown et al. [14] chemically coupled different concentrations of type I collagen with PLGA and established a kind of extracellular matrix porous culture system using wet electrospinning technology. Type I collagen can improve the synthetic function of primary hepatocytes over time. The secretory activity and CYP450 gene transcription activity of primary hepatocytes cultured in this system have been significantly improved, and it can be used as primary hepatocytes for a long time. *Cultivated biosynthetic starting material*: Electrospinning technology can introduce various biologically active macromolecules into the fiber surface to adjust the biological behavior of the carried cells. The abundance and availability of adipose-derived stem cells may prove to be a new cell therapy for bone repair and regeneration. The introduction of poly-L-benzyl glutamate and hydroxyapatite into the surface of electrospun fibers can promote the osteogenic differentiation of adipose-derived stem cells without adding induction medium [15]. The influence of collagen electrospun fibers on cell behavior is not limited to the introduction of various active substances on the surface, and the purpose of regulating the biological behavior of cells can also be achieved through further control of the spinning structure. The PCL/collagen nanofibers prepared by electrospinning technology have fiber orientation, which can well induce the oriented arrangement of human skeletal muscle cells and promote the formation of myotubes [16]. The electrospun nerve catheter blended with caprolactone (ε) and type I collagen has also been proven to effectively promote the recovery of the continuity of the broken-chain nerve [17] (Figure 7.3).

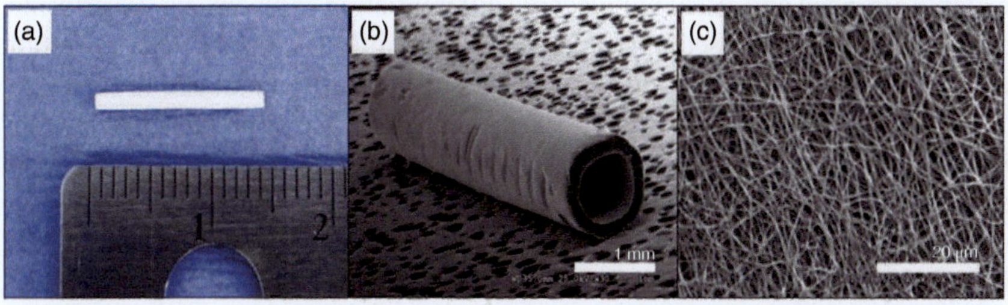

Figure 7.3 (a) The gross appearance and SEM images of electrospun PCL/collagen conduits: (b) entire (×30) and (c) surface (×2.0 K). Source: Ref. [17]/Elsevier.

7.2.2 Tissue Regeneration

Because collagen is unstable in volatile organic solvents, a large amount of gelatin fibers is produced during the spinning process, so there is a huge challenge in preparing ultra-fine collagen fiber structures. Truong et al. [18] reported that alternate coating of triple helix anion and cation collagen layers is an effective method for preparing the same type of collagen coating. This method uses non-denaturing conditions, so collagen maintains its natural triple helix structure and will not be converted into gelatin. The collagen-coated fibers they prepared provide a good matrix for cell adhesion and spreading while maintaining dimensional stability, providing a huge space for the further application of collagen in the field of electrospinning. Collagen is a good raw material for constructing a biomimetic extracellular matrix environment and has extremely low immunogenicity. The mechanical properties and thermal stability of collagen nanofibers are suitable for application on human skin. They can promote the survival of human keratinocytes (HaCaTs) and human dermal fibroblasts (HDFs) and induce epidermal differentiation. It is a star material in the field of skin regeneration [19]. The environment created by collagen that is conducive to cell proliferation will also increase the risk of bacterial and microbial growth. By adding nano-silver particles, the risk of infection caused by tissue engineering scaffold transplantation can be well suppressed [20].

7.3 Collagen-Based 3D Printing for Biomedical Application

7.3.1 Tissue Regeneration

3D printing technology has attracted the attention of researchers in the field of tissue engineering because of its ability to prepare highly personalized scaffold structures. In addition, collagen is a typical extracellular matrix (ECM) bionic material. The combination of the two provides unlimited possibilities for tissue repair and regeneration. In the human body, there are many microtubule structures, such as muscles, blood vessels, nerves, and tendons. Inspired by the structure of lotus and lotus stems, Hwangbo et al. [21] used the sacrificial preparation method of continuously removing the support materials PCL and PVA to successfully construct unidirectionally arranged microtubular collagen scaffold, and has made great progress in inducing the highly arranged and efficient myogenic differentiation of myoblasts. 3D printing technology is also often used in the preparation of multi-layer structure stents, but the general extrusion 3D printing method can only imitate the multi-layer structure or the lack of interconnection channels between layers. These problems restrict its further development. Wu et al. [22] used indirect 3D-TIPS printing technology to develop a three-dimensional biomimetic hybrid composite scaffold as an in vitro airway model. The layered interconnected porous structure of the 3D-TIPS elastic scaffold allows collagen hydrogel to penetrate into the interpenetrating airway. The epithelium on the scaffold is a synergistic bioengineering combination of collagen hydrogel, thermo-responsive elastomer nanohybrid network, epithelial cell culture, and double-layer co-culture, which provides a potential in vitro platform for respiratory disease research and drug screening (Figure 7.4).

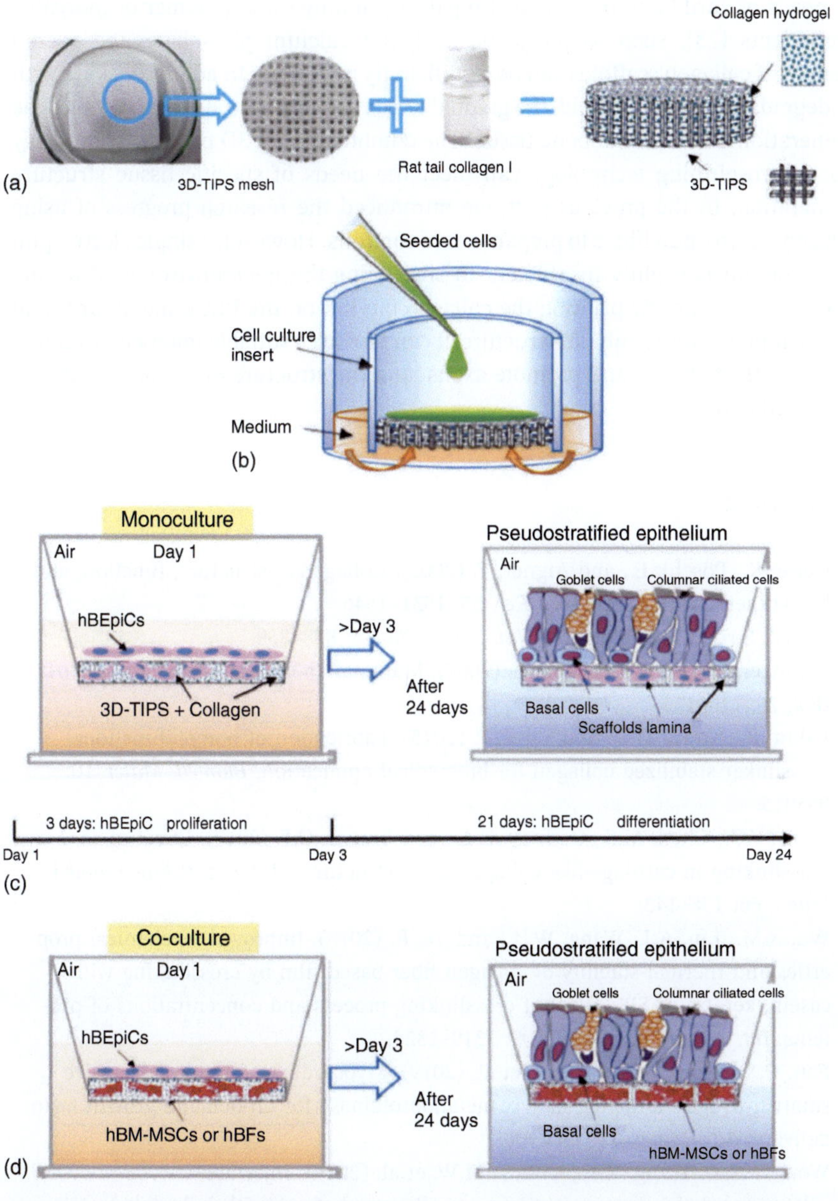

Figure 7.4 (a) Collagen-modified 3D-TIPS elastic scaffolds are described. (b) Schematic diagram of the air–liquid interface cell culture system. (c) Growth of monocultured human bronchial epithelial cells on the scaffolds. (d) Growth of human bronchial epithelial cells co-cultured with stromal cells on the scaffolds. Source: Ref. [22]/Elsevier.

The mechanical properties of pure collagen scaffolds are still unable to meet the needs of hard bone tissue defect repair. By adding other polymer or inorganic components [23], such as polylactic acid and calcium phosphate, the overall strength of collagen scaffolds can be significantly improved. In addition, along with the degradation of the scaffold, the gradual release of mineral ions also promotes the regeneration and repair of bone tissue. The combination of 3D printing technology and electrospinning technology can meet the needs of specific tissue structure regeneration. In the previous part, we introduced the research progress of using collagen electrospun fibers to prepare nerve conduits. However, a single electrospun nerve conduit is slightly insufficient in simulating the microenvironment during nerve regeneration. By printing the collagen bio-ink on the PLCL membrane, and then crimping it into a tubular structure, it can provide a suitable microenvironment for nerve regeneration and promote axons, and the structure of the myelin sheath is restored [24].

References

1 Gelse, K., Pöschl, E., and Aigner, T. (2003). Collagens—structure, function, and biosynthesis. *Adv. Drug Deliv. Rev.* 55: 1531–1546.

2 (1998). Structure of collagen. *Nat. Struct. Biol.* 5: 858–859.

3 Leitinger, B. and Hohenester, E. (2007). Mammalian collagen receptors. *Matrix Biol.* 26: 146–155.

4 Lakra, R., Kiran, M.S., and Sai, K.P. (2015). Fabrication of homobifunctional crosslinker stabilized collagen for biomedical application. *Biomed. Mater.* 10: 065015.

5 Chen, Y.C., Chen, M.S., Gaffney, E.A., and Brown, C.P. (2017). Effect of crosslinking in cartilage-like collagen microstructures. *J. Mech. Behav. Biomed. Mater.* 66: 138–143.

6 Wu, X.M., Liu, A.J., Wang, W.H., and Ye, R. (2018). Improved mechanical properties and thermal-stability of collagen fiber based film by crosslinking with casein, keratin or SPI: effect of crosslinking process and concentrations of proteins. *Int. J. Biol. Macromol.* 109: 1319–1328.

7 Fan, C.X., Shi, J.J., Zhuang, Y. et al. (2019). Myocardial-infarction-responsive smart hydrogels targeting matrix metalloproteinase for on-demand growth factor delivery. *Adv. Mater.* 31: 1902900.

8 Wong, F.S.Y., Tsang, K.K., Chu, A.M.W. et al. (2019). Injectable cell-encapsulating composite alginate-collagen platform with inducible termination switch for safer ocular drug delivery. *Biomaterials* 201: 53–67.

9 Hinman, S.S., Wang, Y.L., Kim, R., and Allbritton, N.L. (2021). In vitro generation of self-renewing human intestinal epithelia over planar and shaped collagen hydrogels. *Nat. Protoc.* 16: 352–382.

10 Zheng, L., Liu, S.J., Cheng, X.J. et al. (2019). Intensified stiffness and photodynamic provocation in a collagen-based composite hydrogel drive chondrogenesis. *Adv. Sci.* 6: 1900099.

11 He, J., Zhang, N.H., Zhu, Y. et al. (2021). MSC spheroids-loaded collagen hydrogels simultaneously promote neuronal differentiation and suppress inflammatory reaction through PI3K-Akt signaling pathway. *Biomaterials* 265: 120448.

12 Mredha, M.T.I., Kitamura, N., Nonoyama, T. et al. (2017). Anisotropic tough double network hydrogel from fish collagen and its spontaneous in vivo bonding to bone. *Biomaterials* 132: 85–95.

13 Chen, F., Le, P., Fernandes-Cunha, G.M. et al. (2020). Bio-orthogonally crosslinked hyaluronate-collagen hydrogel for suture-free corneal defect repair. *Biomaterials* 255: 120176.

14 Brown, J.H., Das, P., DiVito, M.D. et al. (2018). Nanofibrous PLGA electrospun scaffolds modified with type I collagen influence hepatocyte function and support viability. *Acta Biomater.* 73: 217–227.

15 Ravichandran, R., Venugopal, J.R., Sundarrajan, S. et al. (2012). Precipitation of nanohydroxyapatite on PLLA/PBLG/collagen nanofibrous structures for the differentiation of adipose derived stem cells to osteogenic lineage. *Biomaterials* 33: 846–855.

16 Choi, J.S., Lee, S.J., Christ, G.J. et al. (2008). The influence of electrospun aligned poly(ε-caprolactone)/collagen nanofiber meshes on the formation of self-aligned skeletal muscle myotubes. *Biomaterials* 29: 2899–2906.

17 Lee, B.K., Ju, Y.M., Cho, J.G. et al. (2012). End-to-side neurorrhaphy using an electrospun PCL/collagen nerve conduit for complex peripheral motor nerve regeneration. *Biomaterials* 33: 9027–9036.

18 Truong, Y.B., Glattauer, V., Briggs, K.L. et al. (2012). Collagen-based layer-by-layer coating on electrospun polymer scaffolds. *Biomaterials* 33: 9198–9204.

19 Zhou, T., Wang, N.P., Xue, Y. et al. (2015). Development of biomimetic tilapia collagen nanofibers for skin regeneration through inducing keratinocytes differentiation and collagen synthesis of dermal fibroblasts (Retraction of vol 7, pg 3253, 2015). *ACS Appl. Mater. Interfaces* 7: 19863.

20 Qian, Y.Z., Zhou, X.F., Zhang, F.M. et al. (2019). Triple PLGA/PCL scaffold modification including silver impregnation, collagen coating, and electrospinning significantly improve biocompatibility, antimicrobial, and osteogenic properties for orofacial tissue regeneration. *ACS Appl. Mater. Interfaces* 11: 37381–37396.

21 Hwangbo, H., Kim, W., and Kim, G.H. (2021). Lotus-root-like microchanneled collagen scaffold. *ACS Appl. Mater. Interfaces* 13: 12656–12667.

22 Wu, L.X., Magaz, A., Huo, S.G. et al. (2020). Human airway-like multilayered tissue on 3D-TIPS printed thermoresponsive elastomer/collagen hybrid scaffolds. *Acta Biomater.* 113: 177–195.

23 Dewey, M.J., Nosatov, A.V., Subedi, K. et al. (2021). Inclusion of a 3D-printed hyperelastic bone mesh improves mechanical and osteogenic performance of a mineralized collagen scaffold. *Acta Biomater.* 121: 224–236.

24 Yoo, J., Park, J.H., Kwon, Y.W. et al. (2020). Augmented peripheral nerve regeneration through elastic nerve guidance conduits prepared using a porous PLCL membrane with a 3D printed collagen hydrogel. *Biomater. Sci.* 8: 6261–6271.

8

Sources, Structures, and Properties of Gelatin

Gelatin is a kind of protein obtained by partial hydrolysis of collagen. Collagen has a rod-like triple helix structure. When it is partially hydrolyzed to prepare gelatin, the triple helix structure of collagen is partially separated and broken. In the process of hydrolysis, the Arg-Gly-Asp peptide sequence (RGD) and matrix metalloproteinases (MMPs) degradation motif of collagen are preserved, so that gelatin and collagen have similar biological functions [1]. At present, gelatin can be divided into three categories according to different treatment methods: type A gelatin obtained from pigskin treated with acid, type B gelatin obtained from cowhide treated by alkali, and enzymatic gelatin. Different pretreatment will also affect the electrical properties of collagen and produce gelatin with different isoelectric points. In the alkaline process, the amide group in gelatin is hydrolyzed into carboxyl group. The high density of carboxyl group makes gelatin negatively charged, which is convenient for assembly with positively charged and alkaline substances. On the other hand, the acid treatment has little effect on the amide group, which is beneficial to the assembly with acidic substances.

At present, researchers mainly improve the mechanical properties and biological functions of gelatin by means of cross-linking, chemical grafting, and blending. Various gelatin derivatives are widely used in the fields of cell culture, drug delivery, and tissue engineering.

8.1 Gelatin-Based Hydrogel for Biomedical Application

8.1.1 Cell Culture and Behavior Regulation

The regulation of cell behavior by hydrogels has always been a hot research issue in the field of biomedicine. How to better simulate the primitive environment of cell growth has always been the direction of the efforts of the majority of researchers. The PEGX method is a commonly used hydrogel cross-linking strategy, which can generate a stable cross-linked gelatin hydrogel while coupling bioactive peptides to the gelatin polymer. Based on this method, the polyethylene glycol covalently cross-linked gelatin hydrogel material can play a biological activity similar to the cell basement membrane by coupling the laminin-derived YIGSR polypeptide or

Natural Polymers for Biomedical Applications, First Edition. Wenguo Cui and Lei Xiang.

the VEGF similar QK polypeptide, and obviously affect the biological behavior of endothelial cells [2]. The higher-level requirement of cell culture is how to transport stem cells to the lesion site to achieve the therapeutic purpose of tissue regeneration while maintaining efficient differentiation of stem cells. Injectable hydrogel is an effective vehicle for this goal of experimentation.

Previous studies have emphasized the benefits of macroporous scaffolds in enhancing the survival of stem cells in vivo, but the role of such macroporous hydrogels in immunocompetent animal models has not been systematically studied. Tang et al. [3] further optimized the gelatin hydrogel material in response to this problem and confirmed the effectiveness of the material in a mouse bone defect model with normal immune function. They developed an injectable and in situ cross-linkable gelatin microband (μRB)-based macroporous hydrogel through wet spinning technology. By changing the concentration of glutaraldehyde to optimize the internal cross-linking and fibrinogen coating of the μRB shape, the injectability was optimized, and the bone regeneration speed was significantly improved through in situ injection, and no immune stress events were found. In addition, changing the physical properties of the hydrogel can also significantly improve the proliferation and differentiation behavior of cells. Satapathy et al. [4] innovatively combined nano-alkylene oxide-enhanced gelatin water prepared by microplasma-assisted cross-linking method. The gel is used in cartilage tissue engineering. They found that the prepared hydrogel has unique properties, such as medium surface roughness and viscoelasticity that changes with temperature, and found the regeneration of healthy hyaline cartilage in fracture models (Figure 8.1).

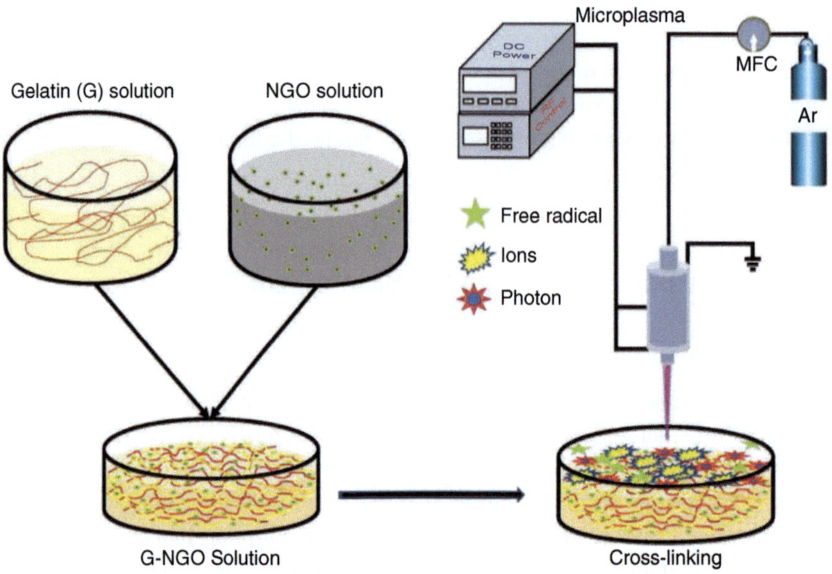

Figure 8.1 Schematic of G-NGO hydrogel synthesized by the Ar-NT microplasma process. Source: Ref. [4]/American Chemical Society.

8.1.2 Drug Delivery

In addition to the method of regulating the biological behavior of cells by loading the cells inside the hydrogel, the hydrogel itself is also an excellent drug delivery vehicle. In view of the potential biological toxicity of commonly used chemical cross-linking methods, GelMA modified with methacrylic anhydride has photocross-linking activity that can significantly reduce the potential biological toxicity risk. In response to the problem of impaired recruitment of mesenchymal stem cells in chronic non-healing wounds, Yu [5] and others used chemokine stromal-cell-derived factor-1α (SDF-1α) to promote the chemotaxis of mesenchymal stem cells. The plastid form is carried in the hydrogel to realize the controllable release of cytokines, and achieve the purpose of promoting the recruitment of mesenchymal stem cells and wound healing in chronic injury sites. In addition, the GelMA hydrogel biomaterial platform can meet the needs of specific tissue repair and regeneration by adding different types of compounds. For example, by incorporating ultraviolet cross-linkable gold nanorods, the hydrogel can have conductive properties, and it can significantly promote the regeneration of myocardial cells and the recovery of cardiac electrophysiological rhythm after myocardial infarction [6].

8.1.3 Tissue Regeneration

The design of microenvironment-responsive hydrogels according to the characteristics of the special microenvironment surrounding pathological tissues is also the current hot research direction. For example, the overexpression of ROS contributes to the pathogenesis of many diseases, such as atherosclerosis, myocardial infarction, and chronic inflammation. Therefore, it is of great significance to develop materials that can locally control the adverse effects caused by excessive ROS production. For this reason, Thi et al. [7] introduced the antioxidant gallic-acid-conjugated gelatin (GGA) into the gelatin-hydroxyphenylpropene acid (GH) hydrogel to create an injectable hydrogel with enhanced free radical scavenging properties compared to pure GH hydrogel. The resulting GH/GGA hydrogel effectively scavenged hydroxyl free radicals and the scavenging ability can be adjusted by changing the GGA concentration to meet the needs of different pathological environments.

8.2 Gelatin-Based Electrospinning for Biomedical Application

8.2.1 Cell Culture

Gelatin electrospun membrane is a good extracellular matrix scaffold, which can provide a good environment for cell proliferation and differentiation. Although the use of pure gelatin scaffold can meet the basic proliferation requirements of cells, it lacks the mechanism to induce its directed differentiation. For this reason, various biologically active molecules are used to modify gelatin electrospun scaffolds, such as the use of poly-L-lactic acid and gelatin electrospun membranes to control the release of retinoic acid and paclitaxel to promote neuronal differentiation [8]; using glycosaminoglycans gelatin modified with quaternary ammonium salt can significantly promote the differentiation of human bone marrow mesenchymal stem cells to epithelial cells [9], and the function of smooth muscle cells can be directly regulated by changing the concentration of transforming growth factor β-2 loaded on the gelatin electrospinning scaffold [10]. In the future, stromal-cell-derived factor-1a was added to promote the in situ recruitment of mesenchymal stem cells of the gelatin electrospinning scaffold [11] and so on.

8.2.2 Tissue Regeneration

The oriented electrospun membrane prepared by electrospinning technology can simulate the natural microtubule-like structure in the body. The concentration of the gelatin solution, the feeding speed, and the electric field potential between the collector plate and the injection needle can be adjusted accurately. For example, gelatin multi-walled carbon nanotubes electrospun scaffolds can form tightly packed myotube-like structures to promote the arrangement and differentiation of myoblasts, thereby generating functional muscle fibers [12]. Using a similar method, the randomly aligned electrospun scaffold prepared from the gelatin electrospinning solution loaded with basic fibroblast growth factor can promote the arrangement of endothelial cells along the fiber and observe the development of filamentous feet, which can promote the capillary revascularization is of great significance [13].

The anisotropic electrospinning scaffold prepared based on the gelatin component also has great application value in the field of cardiac tissue engineering. Kharaziha et al. [14] prepared a nanofiber scaffold with clear anisotropy that mimics the myocardial structure of the left ventricle by incorporating sebacic acid glycerol-sebacate (PGS). They found that the nanofiber scaffold composed of 33 wt% PGS induced the best synchronized contraction of CMS, and at the same time significantly enhanced the arrangement of cells, because the presence of PGS promoted the overall formation of sarcomeres in the arrangement and random structure. In addition, adhesion of the sternum and epicardium after cardiac surgery increases the risk and complexity of reoperation of the heart, and is an important cause of poor prognosis. The bioabsorbable gelatin/PCL composite membrane prepared by the electrospinning method can effectively inhibit cell infiltration, resist the adhesion of the sternum and the epicardium, and it will spontaneously degrade and avoid the risk of foreign bodies remaining in the body [15].

How to reduce the biological toxicity of cross-linking agents is also the main research direction in the field of electrospinning. In addition to the method of modifying gelatin with methacrylic anhydride and using photocross-linking mentioned in the previous section, it is also possible to use low-toxicity chemical cross-linking agents. At present, genipin is widely used as a low-toxicity cross-linking agent in the biomedical field. For example, genipin cross-linked gelatin electrospun scaffolds are used as the activator of decellularized rat brain extracellular matrix. The method can provide basic biochemical signals for seed cells and induce stromal cells to differentiate into neural precursor cells [16]. The concentration of the added genipin also affects the drug release rate of the gelatin electrospun stent. The VEGF cumulative release rate of the 0.1% w/v genipin cross-linked gelatin stent is 1.5 times that of 0.5% w/v [17].

8.3 Gelatin-Based 3D Printing for Biomedical Application

8.3.1 Tissue Regeneration

3D bioprinting can be an advanced process that combines materials and cells to meet the special needs of various tissue engineering. The bio-ink used for 3D printing must have good rheological properties and can quickly form a stable hydrogel in a cell-compatible manner. For this reason, gelatin is usually modified with acryl functional groups. As a thiol-ene clickable bio-ink, allyl gelatin has great application potential in the field of biomedicine. The curing of this system is carried out by dimerization, which can produce a network with flexible properties. Compared with free radical polymerization, the cross-linking process of this bio-ink does not produce additional non-degradable components [18]. In addition to the inherent properties of bio-inks that will affect cell culture, many parameters in the 3D printing process, such as bio-ink concentration, printing temperature, printing speed, cell density, scaffold pores, and geometric structure, directly affect cell growth [19, 20].

Gelatin-based bio-inks often achieve the effect of $1 + 1 > 2$ through ingenious combination with other polymers. Traditional gelatin hydrogels have poor mechanical properties and cannot be used as load-bearing scaffolds. Combining silk fibroin with gelatin can make the mechanical properties and degradation rate of 3D printed scaffolds better meet the needs of cartilage repair in situ. This gelatin–silk fibroin composite bio-ink not only retains sufficient bone marrow stem cell BMSCs to efficiently recruit, but also acts as a physical barrier to block blood clots, provides mechanical protection before the formation of new cartilage, and contributes to the proliferation and growth of BMSCs. Differentiation and the production of extracellular matrix provide a suitable three-dimensional microenvironment [21]. In addition, through the introduction of cleavable poly(*N*-acryl-2-glycine) (PACG) through the enhanced hydrogen bonding with GelMA, the 3D printed scaffold has high tensile strength (up to 1.1 MPa) and excellent compressive strength (up to 12.4 MPa), large Young's modulus (up to 320 kPa), and high compression modulus (up to 837 kPa). At the same time, the cross-linking of GelMA can stabilize the temporary PACG network, and the biodegradation rate of the material can be controlled by adjusting the ratio of ACG/GelMA [22] (Figure 8.2).

At present, gelatin bio-inks still have some problems when applied to 3D printing. For example, low-concentration gelatin bio-inks always have insufficient cross-linking strength, which leads to the collapse of 3D structures and low precision in the preparation of small structures (such as capillaries). By adding 2,2,6,6-tetramethylpiperidine-1-oxyl radical oxidized cellulose nanofiber can solve

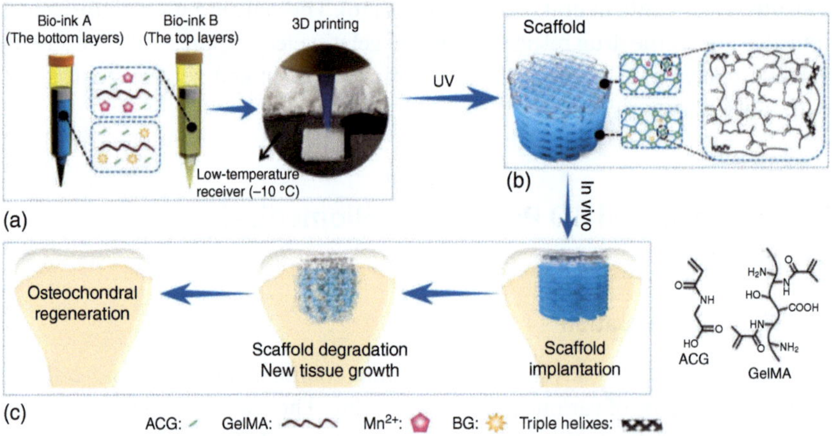

Figure 8.2 (a) The composition of two bioinks and the method for 3D printing of gradient scaffolds are described. (b) UV cross-linking to form a stable hydrogel network and major hydrogen-bonding interactions. (c) Application of gradient hydrogel scaffolds containing different compositions for repairing osteochondral defects in an animal model. Source: Ref. [22]/John Wiley & Sons/CC BY 4.0.

the problem of cross-link density. In addition, the presence of GelMA improves the rheological properties of cellulose nanofiber and further expands the application range of low-concentration gelatin bio-inks in the field of 3D printing [23]. Li et al. [24] increased the overall strength of the material by adding carbon nanotubes to the gelatin bio-ink, and then used the vertical oriented extrusion of the printing nozzle and the axial rotation of the stepper motor assembly to successfully prepare the inner diameter. The microvascular structure with a thickness of 3 mm, an average wall thickness of 0.5 mm, and a length of 7–10 cm realizes the precise preparation of the microstructure.

References

1 Ludwig, P.E., Huff, T.J., and Zuniga, J.M. (2018). The potential role of bioengineering and three-dimensional printing in curing global corneal blindness. *J. Tissue Eng.* 9: https://doi.org/10.1177/2041731418769863.

2 Su, J., Satchell, S.C., Wertheim, J.A., and Shah, R.N. (2019). Poly(ethylene glycol)-crosslinked gelatin hydrogel substrates with conjugated bioactive peptides influence endothelial cell behavior. *Biomaterials* 201: 99–112.

3 Tang, Y.H., Tong, X.M., Conrad, B., and Yang, F. (2020). Injectable and in situ crosslinkable gelatin microribbon hydrogels for stem cell delivery and bone regeneration. *Theranostics* 10: 6035–6047.

4 Satapathy, M.K., Manga, Y.B., Ostrikov, K.K. et al. (2020). Microplasma cross-linked graphene oxide-gelatin hydrogel for cartilage reconstructive surgery. *ACS Appl. Mater. Interfaces* 12: 86–95.

5 Yu, J.R., Janssen, M., Liang, B.J. et al. (2020). A liposome/gelatin methacrylate nanocomposite hydrogel system for delivery of stromal cell-derived factor-1α and stimulation of cell migration. *Acta Biomater.* 108: 67–76.
6 Navaei, A., Saini, H., Christenson, W. et al. (2016). Gold nanorod-incorporated gelatin-based conductive hydrogels for engineering cardiac tissue constructs. *Acta Biomater.* 41: 133–146.
7 Thi, P.L., Lee, Y., Tran, D.L. et al. (2020). forming and reactive oxygen species-scavenging gelatin hydrogels for enhancing wound healing efficacy. *Acta Biomater.* 103: 142–152.
8 Binan, L., Tendey, C., De Crescenzo, G. et al. (2014). Differentiation of neuronal stem cells into motor neurons using electrospun poly-L-lactic acid/gelatin scaffold. *Biomaterials* 35: 664–674.
9 Bhowmick, S., Scharnweber, D., and Koul, V. (2016). Co-cultivation of keratinocyte-human mesenchymal stem cell (hMSC) on sericin loaded electrospun nanofibrous composite scaffold (cationic gelatin/hyaluronan/chondroitin sulfate) stimulates epithelial differentiation in hMSCs: in vitro study. *Biomaterials* 88: 83–96.
10 Ardila, D.C., Tamimi, E., Danford, F.L. et al. (2015). TGFβ2 differentially modulates smooth muscle cell proliferation and migration in electrospun gelatin-fibrinogen constructs. *Biomaterials* 37: 164–173.
11 Ji, W., Yang, F., Ma, J.L. et al. (2013). Incorporation of stromal cell-derived factor-1α in PCL/gelatin electrospun membranes for guided bone regeneration. *Biomaterials* 34: 735–745.
12 Ostrovidov, S., Shi, X.T., Zhang, L. et al. (2014). Myotube formation on gelatin nanofibers – multi-walled carbon nanotubes hybrid scaffolds. *Biomaterials* 35: 6268–6277.
13 Montero, R.B., Vial, X., Nguyen, D.T. et al. (2012). bFGF-containing electrospun gelatin scaffolds with controlled nano-architectural features for directed angiogenesis. *Acta Biomater.* 8: 1778–1791.
14 Kharaziha, M., Nikkhah, M., Shin, S.R. et al. (2013). PGS:Gelatin nanofibrous scaffolds with tunable mechanical and structural properties for engineering cardiac tissues. *Biomaterials* 34: 6355–6366.
15 Feng, B., Wang, S.B., Hu, D.J. et al. (2019). Bioresorbable electrospun gelatin/polycaprolactone nanofibrous membrane as a barrier to prevent cardiac postoperative adhesion. *Acta Biomater.* 83: 211–220.
16 Baiguera, S., Del Gaudio, C., Lucatelli, E. et al. (2014). Electrospun gelatin scaffolds incorporating rat decellularized brain extracellular matrix for neural tissue engineering. *Biomaterials* 35: 1205–1214.
17 Del Gaudio, C., Baiguera, S., Boieri, M. et al. (2013). Induction of angiogenesis using VEGF releasing genipin-crosslinked electrospun gelatin mats. *Biomaterials* 34: 7754–7765.
18 Bertlein, S., Brown, G., Lim, K.S. et al. (2017). Thiol–ene clickable gelatin: a platform bioink for multiple 3D biofabrication technologies. *Adv. Mater.* 29: 1703404.

19 Lewis, P.L., Green, R.M., and Shah, R.N. (2018). 3D-printed gelatin scaffolds of differing pore geometry modulate hepatocyte function and gene expression. *Acta Biomater.* 69: 63–70.

20 Billiet, T., Gevaert, E., De Schryver, T. et al. (2014). The 3D printing of gelatin methacrylamide cell-laden tissue-engineered constructs with high cell viability. *Biomaterials* 35: 49–62.

21 Shi, W.L., Sun, M.Y., Hu, X.Q. et al. (2017). Structurally and functionally optimized silk-fibroin–gelatin scaffold using 3D printing to repair cartilage injury in vitro and in vivo. *Adv. Mater.* 29: 1701089.

22 Gao, F., Xu, Z.Y., Liang, Q.F. et al. (2019). Osteochondral regeneration with 3D-printed biodegradable high-strength supramolecular polymer reinforced-gelatin hydrogel scaffolds. *Adv. Sci.* 6: 1900867.

23 Xu, W.Y., Molino, B.Z., Cheng, F. et al. (2019). On low-concentration inks formulated by nanocellulose assisted with gelatin methacrylate (GelMA) for 3D printing toward wound healing application. *ACS Appl. Mater. Interfaces* 11: 8838–8848.

24 Li, L.Y., Qin, S., Peng, J. et al. (2020). Engineering gelatin-based alginate/carbon nanotubes blend bioink for direct 3D printing of vessel constructs. *Int. J. Biol. Macromol.* 145: 262–271.

9

Sources, Structures, and Properties of Silk Fibroin

Silk fibroin is a natural high-molecular-weight fibrin secreted from the silk glands of silkworms, spiders, and other arthropods. Human application of silk fibroin can be traced back to thousands of years ago. Silk fibroin consists of light chain and heavy chain. The highly conserved (GAGAGS) and the less conserved (GAGX) repeat sequence (where X represents valine or tyrosine) on the heavy chain determine the β-sheet conformation of silk fibroin which is very similar to type I collagen [1].

Silk fibroin has good mechanical and physical and chemical properties, and different forms can be obtained after modification, such as fiber, membrane, and gel. At present, most of the silk fibroin used in biological materials is from silkworm. Spider silk has better mechanical properties, but its difficulty in obtaining restricts its further application. In recent years, MaSp1/MaSp2 artificial spider silk fiber produced by recombinant protein technology has proved that artificially changing the protein sequence composition can change its mechanical properties, which makes silk fibroin coruscate new vitality in the biomedical field.

9.1 Silk-Fibroin-Based Hydrogel for Biomedical Application

9.1.1 Drug Delivery and Cell Culture

Using silk fibroin hydrogel as a carrier to load cells and/or growth factors is a common strategy for bone defect repair. How to achieve the continuous delivery of drugs and cells has always been a difficult problem that researchers want to overcome. By regulating the inner β-sheet (crystalline) density of the silk fibroin hydrogel, not only the mechanical properties of the material can be controlled, but also the anti-hydration performance of the overall hydrogel can be adjusted through the hydrophobic/hydrophilic silk protein–antibody interaction. Control and gel hydration resistance to achieve sustained release of protein drugs [2]. In addition, using nano-bioglass as a cell or growth factor stabilizer can achieve the goal of sustained release. In addition, as the degradation of nano-bioglass leads to the release of related ions, it will further induce cell proliferation and differentiation [3]. In terms of disease treatment, researchers are increasingly turning their attention to

Natural Polymers for Biomedical Applications, First Edition. Wenguo Cui and Lei Xiang.

the study of non-drug treatment methods. At present, hyperthermia is widely used in tumor ablation, by confining the hydrophilic iron oxide nanocubes stabilized by polyethylene glycol to promote ferrimagnetism in the polymer matrix of injectable silk fibroin. The prepared injectable silk fibroin hydrogel with magnetocaloric effect can effectively kill tumor cells lurking deep in the body in an alternating magnetic field [4].

According to different sources, silk fibroin can be divided into two types: mulberry silkworm (*Bombyx mori*) and non-mulberry silkworm (*Antheraea assamensis*) silk, and their role in promoting cell proliferation and differentiation is also distinguished. For example, in terms of chondrocyte culture, non-mulberry silk fibroin–agarose hydrogel showed higher sGAG and collagen content (about 1.5 times) than silkworm–agarose hydrogel. Histological and immunohistochemical analyses further confirmed the increase in the deposition of sGAG and collagen. Various experimental results indicate that different sources of biological materials need to be selected according to different cell types to achieve the best experimental and therapeutic effects [5].

9.1.2 Tissue Regeneration

The application of hydrogel materials in the field of tissue engineering has a history of decades, and whether the mechanical strength of hydrogel materials can match tissues or organs is a constant research hotspot. In the field of tissue engineering, the combined application of multiple processing technologies provides a broad platform for the creation of new repair materials. The high-concentration silk fibroin (16 wt%) hydrogel prepared by salt immersion and freeze-drying technology can ensure the stability of the hydrogel. At the same time, it achieves a higher porosity ($89.3 \pm 0.6\%$) and a wider pore size distribution [6].

At present, improving the mechanical properties of silk fibroin hydrogels is mainly achieved by inducing the transformation of silk fibroin molecules from random

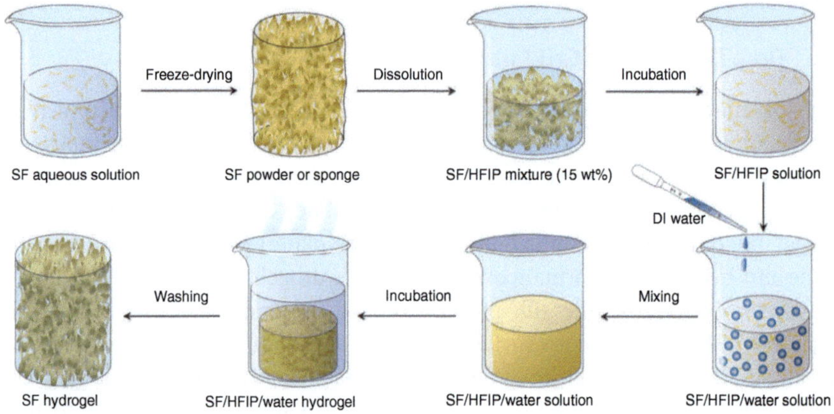

Figure 9.1 Scheme of the BSICT strategy to fabricate pristine SF hydrogels with high mechanical performance. Source: Ref. [8]/John Wiley & Sons.

coils to β-sheet structures. For example, by adding small-peptide gel agents (such as NapFF) to trigger the orderly arrangement of silk fibroin through hydrophobic and hydrogen bond interactions [7], or through moderate binary solvent diffusion and SF/solvent interaction to regulate the gelation process [8], or pre-cross-linking in an alcohol solution to make small β-sheet domains evenly distributed inside the hydrogel [9]. In addition, by improving the types of photoinitiators (such as using riboflavin as the photoinitiator) [10], preparing multi-component hydrogels [11], introducing hydrophobic sacrificial bonds [12] and other methods can be prepared with Silk fibroin hydrogel material with good mechanical strength or elasticity (Figure 9.1).

9.2 Silk-Fibroin-Based Electrospinning for Biomedical Application

9.2.1 Drug Delivery and Antibacterial

Silk fibroin is one of the first natural raw materials used in the textile industry, and was used to make luxurious clothes thousands of years ago. With the emergence of electrospinning technology, silk fibroin is no longer just a material for making clothes, but various applications in the biomedical field have also emerged. A single silk fibroin electrospun fiber cannot meet the needs of a wide variety of biomedicine, which requires that the silk fibroin electrospun fiber must be modified to give it additional biological activity or better mechanical properties. At present, the commonly used modification methods can be divided into two categories: one is to modify the surface of silk fibroin electrospinning fiber, and the other is to modify the components of the silk fibroin electrospinning solution.

Before electrospinning, silk fibroin is used to coat drugs (or other biologically active substances), and the drug release kinetics can be controlled and optimized by adjusting the ratio of silk fibroin and drugs [13, 14]. The addition of environmentally responsive compounds to the silk fibroin electrospinning solution is also a research hotspot this year. For example, by adding nitrous acid diisocyanate to the hybrid nanodiamond-silk fibroin, the thermal stability of the silk fibroin fiber can be improved. The negatively charged nitrogen vacancies (NV-)-color centers embedded in the silk fibers show good fluorescence, which can realize real-time monitoring of wound skin temperature [15].

Surface modification of silk fibroin fibers can give nanofibers more special properties. For example, dopamine coated on the surface of silk fibroin can provide fiber adhesion while stabilizing the additional biologically active substances added to the fiber surface [15]. Using the principles of EDC/NHS and thiol-maleimide click chemistry, peptide sequences (such as the grafted antimicrobial peptide motif Cys-KR12) can be grafted onto the fiber surface to exert antibacterial and other biological activities [16]. In addition, using pyridine solution and chlorosulfonic acid can sulfate the surface of silk fibroin to exert anticoagulant activity [17].

9.2.2 Tissue Regeneration

The optimization of the components of the electrospinning solution can directly affect the structure and properties of the prepared electrospinning silk fibroin scaffold. By adding inorganic composite materials (such as bioglass), the prepared electrospun scaffold can have a hierarchical structure from nanometer to micrometer to better simulate the extracellular matrix [18]. Combining silk fibroin with ε-polycaprolactone can greatly improve the mechanical strength of the prepared scaffold and meet the needs of bone defect repair [19]. Through layer-by-layer coating technology, a variety of natural polymers (such as chitosan and silk fibroin) can be complementarily integrated into one, providing more abundant physical and chemical clues for cell proliferation and differentiation [20].

In addition to various chemical modifications around the silk fibroin body or changing the components of the electrospinning solution to achieve cell behavior control and tissue repair purposes, the mechanical properties of the silk fibroin electrospun fiber and the surface morphology of the prepared electrospun membrane affect the cell. The differentiation also has a direct impact [21], and this mutual relationship urgently needs further elucidation by researchers.

9.3 Silk-Fibroin-Based 3D Printing for Biomedical Application

9.3.1 Tissue Regeneration

There are many forms of 3D printing. The current mainstream printing technologies can be divided into inkjet printing, extrusion printing, and photo-assisted bio-printing, including digital light processing and laser-based printing. How to prepare bio-inks with excellent mechanical and rheological properties is the basis for the further development of 3D printing technology, and how to improve the bioactive functions of bio-inks themselves undoubtedly puts forward a higher demand for the development of new inks.

Tissue engineering scaffolds with rich pores can be prepared by adjusting the printing structure parameters or supplemented by other pore-forming methods (such as freeze-drying technology) to meet the needs of nutrient exchange. However, compared with traditional cast solid bone substitutes, porous bone substitutes have a larger surface area and therefore have a greater risk of postoperative infection. The electrophoretic deposition technique can be used to prepare a silk fibroin coating containing gentamicin on the surface of the porous scaffold to reduce the risk of infection [22, 23].

The mixing of a variety of different components is a common way to modify bio-inks. For example, the mixing ratio of silk fibroin and gelatin or polycaprolactone greatly balances the mechanical properties and degradation rate of 3D printed scaffolds, and can act as a physical barrier to blood clots during the cartilage repair process to form and form new cartilage. The front provides mechanical protection [24, 25]. The combination of silk fibroin, gelatin, hyaluronic acid, and

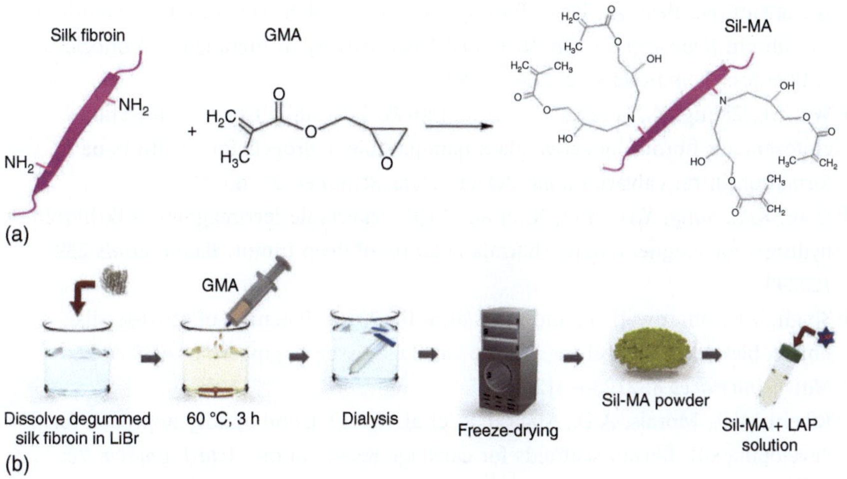

Figure 9.2 (a) Modification of silk protein (SF) molecules with glycerol dimethylacrylate (GMA), which is covalently attached to SF as a double-bond cross-linking site. (b) Synthesis route of methacrylated silk protein. SF was dissolved first, then GMA was added for the modification reaction, followed by dialysis and desalting and freeze-drying, and finally the photoinitiator was added. Source: Ref. [28]/Springer Nature/CC BY 4.0.

tricalcium phosphate can increase the active sites of the bio-ink and optimize its mechanical structure. PRP processing 3D printed scaffolds can significantly promote the cell growth and proliferation of HADMSC [26]. In addition to basic physical and chemical properties, bio-inks also need to be modified according to the selected printing technology. For example, digital light printing technology requires bio-inks to have photocross-linkable properties, while silk fibroin itself does not have photopolymerization cross-linking sites. The technical approach of modifying silk fibroin through glycidyl methacrylate can well meet the printing requirements of digital light printing technology, and by controlling the concentration of silk fibroin bio-ink, the mechanical properties and rheology of the hydrogel can be well adjusted, enabling it to build highly complex organ structures [27, 28] (Figure 9.2).

Recent advances in free-form printing have used the physical properties of microparticle-based granular gels as a medium for immersing bio-inks. However, most of these technologies require post-processing or cross-linking to remove the printed structure, which affects the printing accuracy of complex structures. A new method of gelatinizing silk fibroin step by step in synthetic nanoclay (Laponite) and PEG suspensions can gelate β flakes and silk into arbitrary knots to prepare organ structures of arbitrary geometric shapes [29].

References

1 Bini, E., Knight, D.P., and Kaplan, D.L. (2004). Mapping domain structures in silks from insects and spiders related to protein assembly. *J. Mol. Biol.* 335: 27–40.

2 Guziewicz, N., Best, A., Perez-Ramirez, B., and Kaplan, D.L. (2011). Lyophilized silk fibroin hydrogels for the sustained local delivery of therapeutic monoclonal antibodies. *Biomaterials* 32: 2642–2650.
3 Wu, J.J., Zheng, K., Huang, X.T. et al. (2019). Thermally triggered injectable chitosan/silk fibroin/bioactive glass nanoparticle hydrogels for in-situ bone formation in rat calvarial bone defects. *Acta Biomater.* 91: 60–71.
4 Qian, K.Y., Song, Y.H., Yan, X. et al. (2020). Injectable ferrimagnetic silk fibroin hydrogel for magnetic hyperthermia ablation of deep tumor. *Biomaterials* 259: 120299.
5 Singh, Y.P., Bhardwaj, N., and Mandal, B.B. (2016). Potential of agarose/silk fibroin blended hydrogel for in vitro cartilage tissue engineering. *ACS Appl. Mater. Interfaces* 8: 21236–21249.
6 Ribeiro, V.P., Morais, A.D., Maia, F.R. et al. (2018). Combinatory approach for developing silk fibroin scaffolds for cartilage regeneration. *Acta Biomater.* 72: 167–181.
7 Cheng, B.C., Yan, Y.F., Qi, J.J. et al. (2018). Cooperative assembly of a peptide gelator and silk fibroin afford an injectable hydrogel for tissue engineering. *ACS Appl. Mater. Interfaces* 10: 12474–12484.
8 Zhu, Z.H., Ling, S.J., Yeo, J.J. et al. (2018). High-strength, durable all-silk fibroin hydrogels with versatile processability toward multifunctional applications. *Adv. Funct. Mater.* 28: 1704757.
9 Su, D.H., Yao, M., Liu, J. et al. (2017). Enhancing mechanical properties of silk fibroin hydrogel through restricting the growth of β-sheet domains. *ACS Appl. Mater. Interfaces* 9: 17490–17499.
10 Applegate, M.B., Partlow, B.P., Coburn, J. et al. (2016). Photocrosslinking of silk fibroin using riboflavin for ocular prostheses. *Adv. Mater.* 28: 2417–2420.
11 Buitrago, J.O., Patel, K.D., El-Fiqi, A. et al. (2018). Silk fibroin/collagen protein hybrid cell-encapsulating hydrogels with tunable gelation and improved physical and biological properties. *Acta Biomater.* 69: 218–233.
12 Meng, L., Shao, C.Y., Cui, C. et al. (2020). Autonomous self-healing silk fibroin injectable hydrogels formed via surfactant-free hydrophobic association. *ACS Appl. Mater. Interfaces* 12: 1628–1639.
13 Chouhan, D., Chakraborty, B., Nandi, S.K., and Mandal, B.B. (2017). Role of non-mulberry silk fibroin in deposition and regulation of extracellular matrix towards accelerated wound healing. *Acta Biomater.* 48: 157–174.
14 Steffi, C., Wang, D., Kong, C.H. et al. (2018). Estradiol-loaded poly(ε-caprolactone)/silk fibroin electrospun microfibers decrease osteoclast activity and retain osteoblast function. *ACS Appl. Mater. Interfaces* 10: 9988–9998.
15 Khalid, A., Bai, D.B., Abraham, A.N. et al. (2020). Electrospun nanodiamond-silk fibroin membranes: a multifunctional platform for biosensing and wound-healing applications. *ACS Appl. Mater. Interfaces* 12: 48408–48419.
16 Song, D.W., Kim, S.H., Kim, H.H. et al. (2016). Multi-biofunction of antimicrobial peptide-immobilized silk fibroin nanofiber membrane: implications for wound healing. *Acta Biomater.* 39: 146–155.

17 Liu, H.F., Li, X.M., Zhou, G. et al. (2011). Electrospun sulfated silk fibroin nanofibrous scaffolds for vascular tissue engineering. *Biomaterials* 32: 3784–3793.

18 Singh, B.N. and Pramanik, K. (2017). Development of novel silk fibroin/polyvinyl alcohol/sol-gel bioactive glass composite matrix by modified layer by layer electrospinning method for bone tissue construct generation. *Biofabrication* 9: 015028.

19 Li, Y., Chen, M., Zhou, W. et al. (2020). Cell-free 3D wet-electrospun PCL/silk fibroin/Sr(2+) scaffold promotes successful total meniscus regeneration in a rabbit model. *Acta Biomater.* 113: 196–209.

20 Chen, J.W., Zhan, Y.F., Wang, Y.B. et al. (2018). Chitosan/silk fibroin modified nanofibrous patches with mesenchymal stem cells prevent heart remodeling post-myocardial infarction in rats. *Acta Biomater.* 80: 154–168.

21 Meinel, A.J., Kubow, K.E., Klotzsch, E. et al. (2009). Optimization strategies for electrospun silk fibroin tissue engineering scaffolds. *Biomaterials* 30: 3058–3067.

22 Han, C.J., Yao, Y., Cheng, X. et al. (2017). Electrophoretic deposition of gentamicin-loaded silk fibroin coatings on 3D-printed porous cobalt-chromium-molybdenum bone substitutes to prevent orthopedic implant infections. *Biomacromolecules* 18: 3776–3787.

23 Bidgoli, M.R., Alemzadeh, I., Tamjid, E. et al. (2019). Fabrication of hierarchically porous silk fibroin-bioactive glass composite scaffold via indirect 3D printing: effect of particle size on physico-mechanical properties and in vitro cellular behavior. *Mater. Sci. Eng., C* 103: 109688.

24 Shi, W.L., Sun, M.Y., Hu, X.Q. et al. (2017). Structurally and functionally optimized silk-fibroin-gelatin scaffold using 3D printing to repair cartilage injury in vitro and in vivo. *Adv. Mater.* 29: 1701089.

25 Li, Z., Wu, N.E., Cheng, J. et al. (2020). Biomechanically, structurally and functionally meticulously tailored polycaprolactone/silk fibroin scaffold for meniscus regeneration. *Theranostics* 10: 5090–5106.

26 Wei, L., Wu, S.H., Kuss, M. et al. (2019). 3D printing of silk fibroin-based hybrid scaffold treated with platelet rich plasma for bone tissue engineering. *Bioact. Mater.* 4: 256–260.

27 Kim, S.H., Seo, Y.B., Yeon, Y.K. et al. (2020). 4D-bioprinted silk hydrogels for tissue engineering. *Biomaterials* 260: 120281.

28 Kim, S.H., Yeon, Y.K., Lee, J.M. et al. (2018). Precisely printable and biocompatible silk fibroin bioink for digital light processing 3D printing (vol 9, 1620, 2018). *Nat. Commun.* 9: 1620.

29 Rodriguez, M.J., Dixon, T.A., Cohen, E. et al. (2018). 3D freeform printing of silk fibroin. *Acta Biomater.* 71: 379–387.

10

Sources, Structures, and Properties of Other Polypeptides

Elastin mainly exists in ligaments and vessel walls. Elastic fibers and collagen fibers coexist, giving the tissue elasticity and tensile strength. Elastin is composed of two types of short peptides arranged alternately. One is a short hydrophobic peptide that imparts elasticity to the molecule; the other is an α-helix rich in alanine and lysine residues, responsible for the formation of cross-links between adjacent molecules. Two forms of elastin have been found: elastin I is present in the nape ligaments, aorta, and skin; elastin II can be obtained from cartilage. Since it mainly exists in the arterial wall, elastin is mostly used for vascular regeneration and remodeling. Through recombinant protein technology, based on retaining the original properties of elastin, the mechanical properties and anti-fatigue properties of elastin are further enhanced [1], further expanding elastin and its derivatives applied in tissue engineering and drug delivery.

Fibrin is a kind of elastic filamentous protein, which is transformed from fibrinogen under the action of thrombin [2]. Fibrin mainly exists in plasma and is closely related to blood coagulation function. Many fibrins bind tightly and provide mechanical strength for individual cells or the entire organism, playing a protective or structural role. In the field of biomedicine, fibrin is often used in the preparation of biodegradable biomaterials through electrospinning, hydrogel, and other technologies, and has a good effect in promoting cell growth and wound healing.

10.1 Other Polypeptides-Based Hydrogel for Biomedical Application

10.1.1 Cell Culture and Delivery

Fibrin plays an important role in peripheral nerve regeneration. In the initial stage of nerve regeneration, longitudinal fibrin cords formed spontaneously by fibrin can induce Schwann cell migration and proliferation as well as axon regeneration. Through the combination of electrospinning and hydrogel, a matrix material with a structure similar to that of natural fibrin can be prepared to induce nerve regeneration and repair [3]. The hypoxic environment after nerve injury inhibits the nerve repair process. By introducing perfluorotributylamine into the fibrin

Natural Polymers for Biomedical Applications, First Edition. Wenguo Cui and Lei Xiang.

hydrogel, continuous oxygen supply to the diseased area can be achieved to reverse the hypoxic environment and accelerate nerve regeneration [4]. Microfluidic or water-in-oil emulsification technology can be used to prepare fibrin hydrogel microspheres with a diameter of hundreds of microns. By directly embedding human endothelial cells and fibroblasts or various types of stem cells in the microbead matrix, targeted cell delivery can be achieved, providing a new technical means for the realization of cell therapy [5, 6]. In addition, fibrin-based hydrogels can control cell differentiation, and promote muscle regeneration, bone and cartilage repair, and vascular regeneration by adjusting their mechanical properties, adding magnetic induction materials or carrying various biologically active substances (such as transforming growth factor TGF-β3). Both have outstanding performance [7–10].

10.1.2 Tissue Engineering and Drug Delivery

Elastin can be used to prepare hydrogels of different properties by chemical cross-linking, enzymatic cross-linking, or photocross-linking, and be used in stem cell delivery, angiogenesis, cartilage repair, and many other fields [11–14]. For example, the thiol in a pair of cysteine residues in the elastin-like peptide sequence can form disulfide bonds under ultraviolet light, thereby forming up to 420% elongation and compressive fatigue resistance [15]. In addition, the injectable hydrogel prepared by the dynamic covalent hydrazone bond between hydrazine-modified elastin-like protein and aldehyde-modified hyaluronic acid has become a therapeutic cargo delivery because it can be injected using minimally invasive surgical procedures [11]. The injectable hydrogel prepared by this method can better meet the needs of in situ drug delivery, but its mechanical properties cannot yet meet the requirements of tissue repair. By introducing *N*-acryloyloxy succinimide monomer, the poly(*N*-isopropylacrylamide-*co*-lactic acid-2-hydroxyethyl methacrylate-*co*-polyethylene two alcohol monomethyl ether methacrylate) temperature-sensitive copolymer is functionalized. Reusing ester groups to covalently graft polymers onto α-elastin with primary amine groups can greatly enhance the stability of the hydrogel structure [16]. In addition, by changing the cross-linking environment (for example, cross-linking in high-pressure CO_2), the mechanical properties of the elastic white hydrogel can also be greatly improved [17] (Figure 10.1).

10.2 Other Polypeptides-Based Electrospinning for Biomedical Application

10.2.1 Drug Delivery

Platelet-rich fibrin (PRF), as a reservoir of various growth factors, plays a vital role in wound healing and tissue engineering. Using PRF as a component prepared by electrospinning can significantly promote the proliferation and mineralization induction of MEC3T3-E1 cells [18]. The platelet lysate is loaded on the fibrin fiber

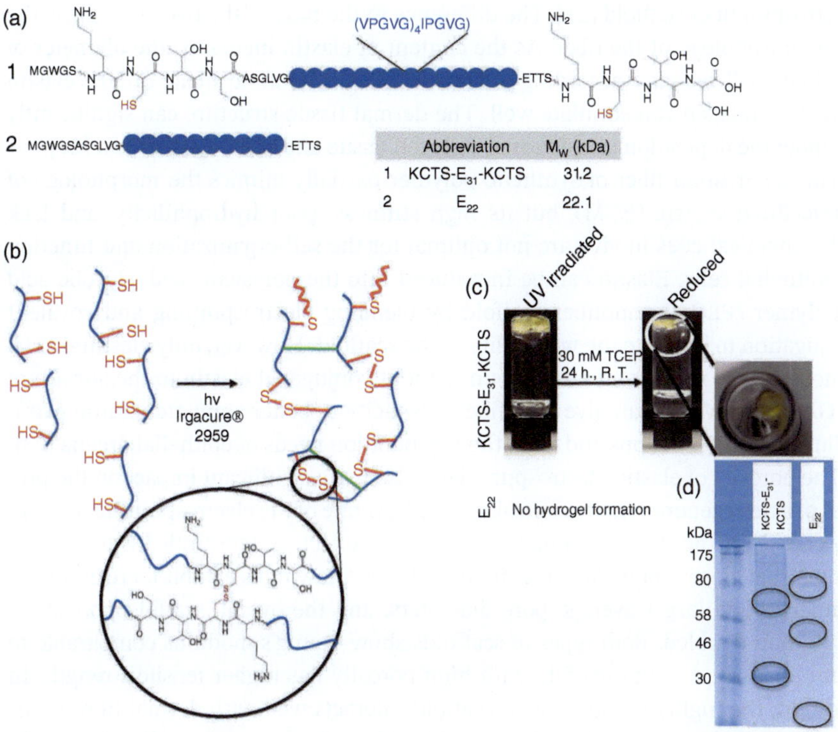

Figure 10.1 (a) Sequences of cysteine-containing ELPs tested for photocross-linking. (b) Mechanism of ELP photocross-linking, including intra-chain and inter-chain cross-linking. (c) Cysteine-containing ELPs were gelatinized by UV irradiation, whereas cysteine-free ELPs were kept in the liquid state. (d) High-molecular-weight bands were shown by electrophoresis after reduction, indicating the formation of bonds other than disulfide. Source: Ref. [15]/John Wiley & Sons.

network by a combination of electrospinning and spray phase change methods. The prepared double-layer fibrin/poly(ether)urethane scaffold can provide sustained release of biologically active platelet-derived growth factors, thereby further accelerating wound healing [19]. In addition, the fibrin-coated nanofiber PLA membrane prepared by needle-free electrospinning technology can provide more mechanical support for the proliferation and diffusion of dermal cells and significantly increase the expression of ECM protein [20].

10.2.2 Tissue Regeneration

In addition, elastin is also the main component of skin maintenance. The blending of collagen and tropoelastin is an effective means to combine the beneficial properties of the two proteins to better mimic the physical, mechanical, and biological clues provided by the dermis of the skin. By adding PEO and NaCl in the spinning process, the homogeneity of the electrospun fiber can be ensured, and then by combining with EDC And NHS cross-linked to prepare a thermally stable electrospun fiber scaffold [21]. The difference in the ratio of the two also affects the basic morphology of the fiber. As the content of elastin increases, the diameter of the fiber will increase accordingly. The electrospun scaffold made of 80% elastin and 20% collagen can simulate well. The dermal tissue structure can significantly promote the deposition of epidermal collagen tissue and skin regeneration [22].

The electrospun fiber of synthetic polymer partially mimics the morphology of extracellular matrix (ECM), but its high stiffness, poor hydrophilicity, and lack of biochemical cues in vivo are not optimal for the self-organization and function of epithelial cells. Elastin can be introduced into the polylactic acid–glycolic acid copolymer (PLGA) nanofiber scaffold by blending electrospinning and covalent conjugation to improve the wettability of the scaffold. However, only elastin–PLGA nanofibers are mixed and electrospun without conjugated elastin to the surface of PLGA fibers, which can give nanofibers elasticity to better simulate natural extracellular matrix junctions and meet the regeneration needs of epithelial organs [23].

The porosity of elastin electrospun fibers also has a significant impact on the process of skin regeneration. By controlling the flow rate of the electrospinning process, the porosity of the electrospinning scaffold can be effectively controlled. Studies have shown that electrospun elastin scaffolds at higher flow rates exhibit increased fiber diameters and larger average pore diameters, and the overall scaffold porosity is more than doubled. Both types of scaffolds show Young's modulus comparable to natural elastin, but the scaffold with high porosity has higher tensile strength. In addition, the highly porous elastin scaffold interacts well with dermal fibroblasts, can promote the colonization of dermal fibroblasts in vitro, and can be tolerated in the subcutaneous implantation model in vivo [24].

In vascular tissues, elastin is an important extracellular matrix protein, which plays an important role in biomechanics and biological signal transduction. Natural elastin is insoluble and difficult to extract from tissues, which results in its relatively small use in the manufacture of vascular tissue engineering scaffolds. The simulated elastin polymer prepared by recombinant technology can show a more complex molecular design and stricter sequence control [25]. Electrospun fibers modified with the surface of recombinant elastin-like polypeptide 4 have been confirmed to promote the proliferation and contraction of smooth muscle cells [26]. In addition, fibrin fiber has good biocompatibility and is an ideal tissue engineering vascular scaffold, but its mechanical properties need to be improved. By mixing polyurethane (PU) or PCL with fibrin, the mechanical properties of fibrin can be significantly enhanced. In vivo experiments have shown that fibrin electrospinning scaffolds can increase the expression of smooth muscle contractile protein and promote ECM deposition and endothelialization of blood vessels [27, 28].

10.3 Other Polypeptides-Based 3D Printing for Biomedical Application

10.3.1 Cell and Organoid Culture

Although three-dimensional (3D) bioprinting is a promising technology to produce tissue-like structures, there is currently a lack of bio-ink configuration strategies that are universally applicable to various types of cells. For this reason, a universal bio-orthogonal bio-ink cross-linking mechanism that is compatible with cells and used with a variety of polymers came into being. In this strategy, gelatin, hyaluronic acid, recombinant elastin-like protein, and PEG are used as backbone polymers to create a storage modulus of 200–10,000 Pa Ink to meet the needs of a variety of cell culture [29]. In addition, the synergistic association of recombinant human tropoelastin and gelatin can achieve high-resolution printing with high cell viability. The vascularized heart construct prepared from this kind of bio-ink realizes the re-function of the endothelial cell barrier and the spontaneous pulsation of cardiomyocytes [30] (Figure 10.2).

10.3.2 Tissue Regeneration

Elastin has long-term stability, elasticity, and biological activity is often used as a matrix material to guide tissue regeneration. By adding gelatin to promote cell adhesion and hyaluronic acid to promote chemical signal transduction between cells to the elastin bio-ink, a new structured soft film with a multilayer structure can be prepared and can well meet the clinical needs of oral epithelial repair [31]. In addition, the bio-ink based on the same composition achieves good cell adhesion and cell proliferation in the reconstruction of amniotic membrane [32]. The preparation of tiny blood vessels has always been a difficult point for 3D printing. The technical method of coating tropoelastin on the surface of the stent and then using ice sacrifice can successfully prepare a layered vascular network [33]. In addition, through the transition of recombinant elastin from disorder to order and self-assembly with graphene, capillary structures with a diameter as low as ~10 μm and a tube wall thickness of ~2 μm have been successfully prepared [34].

The high viscosity of fibrin (cross-linked form) hinders the correct extrusion of the ink, and the fibrinogen in its prepolymer form cannot maintain the fidelity of the shape. This problem can be solved by in situ cross-linking fibrin and blending with other printable polymers [35]. The 3D printed tubular structure can be used not only for vascular regeneration, but also for the repair of other human organs. Compared with blood vessels, the urethra has higher requirements for mechanical properties. The structure and mechanical properties of the natural urethra can be simulated by PCL and PLCL polymers with a spiral structure. Then through bioprinting technology, the polymer scaffold is combined with cell-filled fibrin hydrogel to prepare a tubular structure, which can uniformly deliver urothelial cells and smooth muscle cells to the inner and outer layers of the scaffold, respectively, to realize urethral remodeling [36]. Based on the characteristic that fibrin can induce nerve regeneration well, a 3D printed scaffold with fibrin as the main active substance came into being. Neural cell aggregates based on fibrin 3D printing can significantly cause human-induced pluripotent stem cells to differentiate into neuronal cells and provide a suitable microenvironment for cell growth [37]. In addition, by mixing fibrin with other natural polymers, more targeted cell culture systems can be prepared. For example, fibrin can be added to a photopolymerizable gelatin-based bio-ink mixture to achieve the same mechanical properties as natural myocardial tissue. Cardiac fibroblasts are loaded with targeted bioprinting of constructs to meet the needs of cardiac repair and structural reconstruction [38].

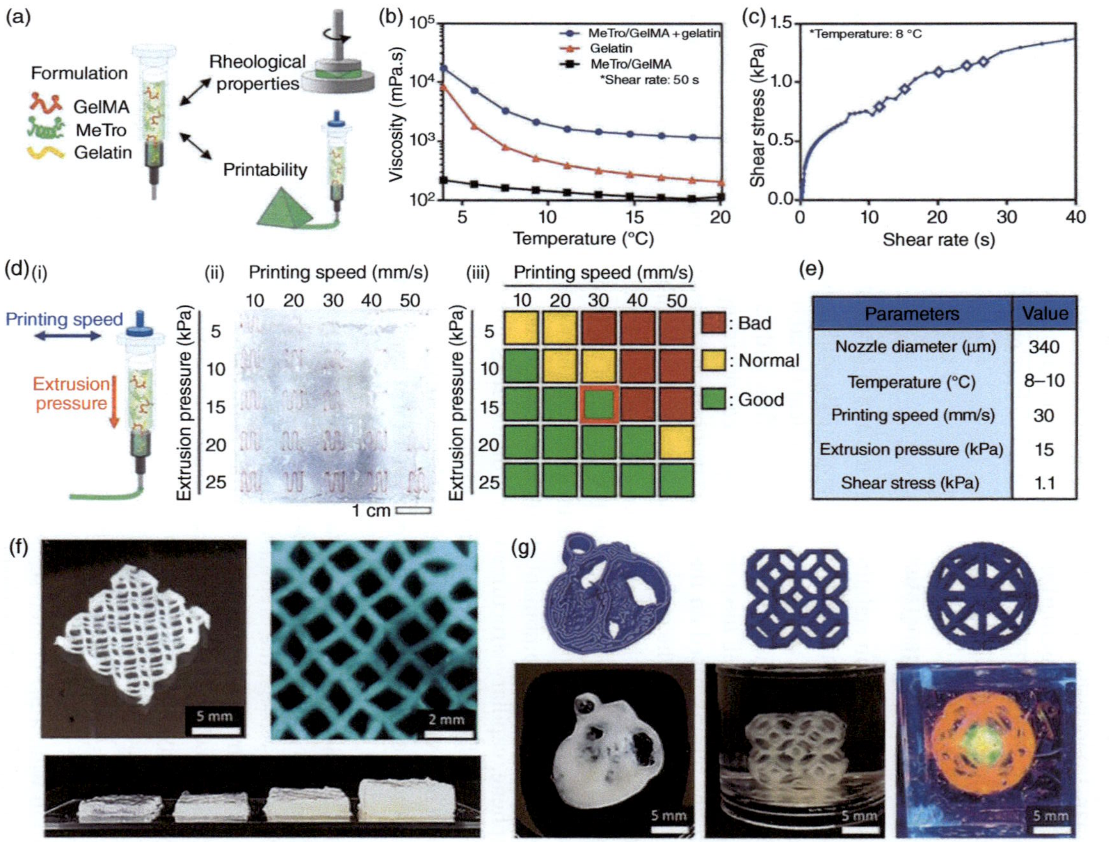

Figure 10.2 (a) Bio-ink formulation. (b) Bio-ink viscosity profile at different temperatures. (c) Bio-ink shear stress profile. (d) Optimization of printing parameters – print speed and pressure. (e) Optimal printing parameters. (f) Printing of multi-layer lattice structures. (g) Printing of various three-dimensional structures. Source: Ref. [30] / John Wiley & Sons.

References

1 Chilkoti, A., Christensen, T., and MacKay, J.A. (2006). Stimulus responsive elastin biopolymers: applications in medicine and biotechnology. *Curr. Opin. Chem. Biol.* 10: 652–657.

2 Bochynska, A.I., Hannink, G., Grijpma, D.W., and Buma, P. (2016). Tissue adhesives for meniscus tear repair: an overview of current advances and prospects for future clinical solutions. *J. Mater. Sci. Mater. Med.* 27: 85.

3 Ma, T., Wang, Y.Q., Qi, F.Y. et al. (2013). The effect of synthetic oxygen carrier-enriched fibrin hydrogel on Schwann cells under hypoxia condition. *Biomaterials* 34: 10016–10027.

4 Du, J.R., Liu, J.H., Yao, S.L. et al. (2017). Prompt peripheral nerve regeneration induced by a hierarchically aligned fibrin nanofiber hydrogel. *Acta Biomater.* 55: 296–309.

5 Rioja, A.Y., Annamalai, R.T., Paris, S. et al. (2016). Endothelial sprouting and network formation in collagen- and fibrin-based modular microbeads. *Acta Biomater.* 29: 33–41.

6 Zhou, H.Z. and Xu, H.H.K. (2011). The fast release of stem cells from alginate-fibrin microbeads in injectable scaffolds for bone tissue engineering. *Biomaterials* 32: 7503–7513.

7 Ho, S.T.B., Cool, S.M., Hui, J.H., and Hutmacher, D.W. (2010). The influence of fibrin based hydrogels on the chondrogenic differentiation of human bone marrow stromal cells. *Biomaterials* 31: 38–47.

8 Marcinczyk, M., Elmashhady, H., Talovic, M. et al. (2017). Laminin-111 enriched fibrin hydrogels for skeletal muscle regeneration. *Biomaterials* 141: 233–242.

9 Deller, R.C., Richardson, T., Richardson, R. et al. (2019). Artificial cell membrane binding thrombin constructs drive in situ fibrin hydrogel formation. *Nat. Commun.* 10: 1887.

10 Arulmoli, J., Wright, H.J., Phan, D.T.T. et al. (2016). Combination scaffolds of salmon fibrin, hyaluronic acid, and laminin for human neural stem cell and vascular tissue engineering. *Acta Biomater.* 43: 122–138.

11 Wang, H.Y., Zhu, D.Q., Paul, A. et al. (2017). Covalently adaptable elastin-like protein-hyaluronic acid (ELP-HA) hybrid hydrogels with secondary thermoresponsive crosslinking for injectable stem cell delivery. *Adv. Funct. Mater.* 27: 1605609.

12 Marsico, G., Jin, C.S., Abbah, S.A. et al. (2021). Elastin-like hydrogel stimulates angiogenesis in a severe model of critical limb ischemia (CLI): an insight into the glyco-host response. *Biomaterials* 269: 120641.

13 Lee, A.J., Lee, Y.J., Jeon, H.Y. et al. (2020). Application of elastin-like biopolymer-conjugated C-peptide hydrogel for systemic long-term delivery against diabetic aortic dysfunction. *Acta Biomater.* 118: 32–43.

14 Staubli, S.M., Cerino, G., De Torre, I.G. et al. (2017). Control of angiogenesis and host response by modulating the cell adhesion properties of an Elastin-Like recombinamer-based hydrogel. *Biomaterials* 135: 30–41.

15 Zhang, Y.N., Avery, R.K., Vallmajo-Martin, Q. et al. (2015). A highly elastic and rapidly crosslinkable elastin-like polypeptide-based hydrogel for biomedical applications. *Adv. Funct. Mater.* 25: 4814–4826.

16 Fathi, A., Mithieux, S.M., Wei, H. et al. (2014). Elastin based cell-laden injectable hydrogels with tunable gelation, mechanical and biodegradation properties. *Biomaterials* 35: 5425–5435.

17 Annabi, N., Mithieux, S.M., Weiss, A.S., and Dehghani, F. (2010). Cross-linked open-pore elastic hydrogels based on tropoelastin, elastin and high pressure CO. *Biomaterials* 31: 1655–1665.

18 Nie, J., Zhang, S.M., Wu, P. et al. (2020). Electrospinning with lyophilized platelet-rich fibrin has the potential to enhance the proliferation and osteogenesis of MC3T3-E1 cells. *Front. Bioeng. Biotechnol.* 8: 595579.

19 Losi, P., Al Kayal, T., Buscemi, M. et al. (2020). Bilayered fibrin-based electrospun-sprayed scaffold loaded with platelet lysate enhances wound healing in a diabetic mouse model. *Nanomaterials (Basel)* 10: 2128.

20 Pajorova, J., Bacakova, M., Musilkova, J. et al. (2018). Morphology of a fibrin nanocoating influences dermal fibroblast behavior. *Int. J. Nanomedicine* 13: 3367–3380.

21 Buttafoco, L., Kolkman, N.G., Engbers-Buijtenhuijs, P. et al. (2006). Electrospinning of collagen and elastin for tissue engineering applications. *Biomaterials* 27: 724–734.

22 Rnjak-Kovacina, J., Wise, S.G., Li, Z. et al. (2012). Electrospun synthetic human elastin:collagen composite scaffolds for dermal tissue engineering. *Acta Biomater.* 8: 3714–3722.

23 Foraida, Z.I., Kamaldinov, T., Nelson, D.A. et al. (2017). Elastin-PLGA hybrid electrospun nanofiber scaffolds for salivary epithelial cell self-organization and polarization. *Acta Biomater.* 62: 116–127.

24 Rnjak-Kovacina, J., Wise, S.G., Li, Z. et al. (2011). Tailoring the porosity and pore size of electrospun synthetic human elastin scaffolds for dermal tissue engineering. *Biomaterials* 32: 6729–6736.

25 Putzu, M., Causa, F., Nele, V. et al. (2016). Elastin-like-recombinamers multilayered nanofibrous scaffolds for cardiovascular applications. *Biofabrication* 8: 045009.

26 Blit, P.H., Battiston, K.G., Yang, M.L. et al. (2012). Electrospun elastin-like polypeptide enriched polyurethanes and their interactions with vascular smooth muscle cells. *Acta Biomater.* 8: 2493–2503.

27 Zhao, L., Li, X.F., Yang, L. et al. (2021). Evaluation of remodeling and regeneration of electrospun PCL/fibrin vascular grafts in vivo. *Mater. Sci. Eng., C* 118: 111441.

28 Yang, L., Li, X.F., Wu, Y.T. et al. (2020). Preparation of PU/fibrin vascular scaffold with good biomechanical properties and evaluation of its performance in vitro and in vivo. *Int. J. Nanomed.* 15: 8697–8715.

29 Hull, S.M., Lindsay, C.D., Brunel, L.G. et al. (2021). 3D bioprinting using UNIversal orthogonal network (UNION) bioinks. *Adv. Funct. Mater.* 31: 2007983.

30 Lee, S., Sani, E.S., Spencer, A.R. et al. (2020). Human-recombinant-elastin-based bioinks for 3D bioprinting of vascularized soft tissues. *Adv. Mater.* 32: e2003915.

31 Tayebi, L., Rasoulianboroujeni, M., Moharamzadeh, K. et al. (2018). 3D-printed membrane for guided tissue regeneration. *Mater. Sci. Eng., C* 84: 148–158.

32 Dehghani, S., Rasoulianboroujeni, M., Ghasemi, H. et al. (2018). 3D-Printed membrane as an alternative to amniotic membrane for ocular surface/conjunctival defect reconstruction: an in vitro & in vivo study. *Biomaterials* 174: 95–112.

33 Wang, R., Ozsvar, J., Aghaei-Ghareh-Bolagh, B. et al. (2019). Freestanding hierarchical vascular structures engineered from ice. *Biomaterials* 192: 334–345.

34 Wu, Y.H., Fortunato, G.M., Okesola, B.O. et al. (2021). An interfacial self-assembling bioink for the manufacturing of capillary-like structures with tuneable and anisotropic permeability. *Biofabrication* 13: 035027.

35 de Melo, B.A.G., Jodat, Y.A., Cruz, E.M. et al. (2020). Strategies to use fibrinogen as bioink for 3D bioprinting fibrin-based soft and hard tissues. *Acta Biomater.* 117: 60–76.

36 Zhang, K., Fu, Q., Atala, A., and Zhao, W. (2017). 3D bioprinting of urethra with PCL/PLCL blend and dual autologous cells in fibrin hydrogel: an in vitro evaluation of biomimetic mechanical property and cell growth environment. *Int. J. Urol.* 24: 16.

37 Abelseth, E., Abelseth, L., De la Vega, L. et al. (2019). 3D printing of neural tissues derived from human induced pluripotent stem cells using a fibrin-based bioink. *ACS Biomater. Sci. Eng.* 5: 234–243.

38 Kumar, S.A., Alonzo, M., Allen, S.C. et al. (2019). A visible light-cross-linkable, fibrin-gelatin-based bioprinted construct with human cardiomyocytes and fibroblasts. *ACS Biomater. Sci. Eng.* 5: 4551–4563.

11

Summary

Polypeptides, including collagen, gelatin, silk fibroin, elastin, and fibrin, are widely distributed in animals. Compared with synthetic polymers without functional groups, the presence of amino and carboxyl groups enables proteins to carry different charges. Therefore, proteins can be easily regulated by pH to form intelligent drug delivery systems. Collagen and its hydrolysate gelatin are the main components of extracellular matrix, which are mostly used in the field of cell culture and tissue engineering in the form of hydrogel and 3D printing. In view of the flexible elasticity or super mechanical properties of silk fibroin, elastin, and fibrin, they are mostly used in electrospinning.

Natural Polymers for Biomedical Applications, First Edition. Wenguo Cui and Lei Xiang.

Section IV

Other Kinds of Natural Polymers for Biomedical Application

Natural Polymers for Biomedical Applications, First Edition. Wenguo Cui and Lei Xiang.

12

Sources, Structures, and Properties of Catechins

Catechin, also known as Caredhieacid, is a kind of phenolic active substance extracted from tea and other natural plants. Catechins have the structure of 2-phenylbenzo-dihydropyran and belong to flavanols with the molecular formula of $C_{15}H_{14}O_6$. Catechin can be subdivided into eight active monomers: catechin, epicatechin (EC), gallocatechin (GC), epigallocatechin (EGC), catechin gallate (CG), epicatechin gallate (ECG), gallocatechin gallate (GCG), and epigallocatechin gallate (EGCG) [1].

The free-radical-scavenging and metal-ion-chelating properties of catechin endow it with a strong antioxidant capacity, among which the antioxidant capacity of EGCG is the best [2, 3]. Some studies have shown that ECG and EGCG can promote muscle regeneration through Myf5-mediated signaling pathways [4] and also reduce bone resorption by down-regulating RANKL nuclear expression [5]. Through the artificial modification of catechin polyhydroxyl structure, catechin can also obtain different pharmacological effects. At present, biomaterials with catechin as the main active component are widely used in the field of drug delivery and tissue engineering.

12.1 Catechins-Based Hydrogel for Biomedical Application

Catechin is the main polyphenolic compound in green tea and has received extensive attention due to its pharmacological effects including antioxidants, antibacterial agents, and antitumor activities. How to improve the bioavailability of catechins has always been a hot research issue. Encapsulating catechins with hydrogel materials is a common method to maintain their stable release. For example, the emulsion is prepared by mixing alginate, oil, and oleate as emulsifiers in sequence. The catechin-loaded calcium alginate microbeads and microparticles prepared by emulsion gel technology can promote drug release as the pH value and incubation temperature increase [6]. The encapsulation efficiency of calcium pectin gel beads that can be loaded with catechins prepared by the inner gel method is greatly improved compared with the microbeads prepared by the outer gel method [7]. In addition, the 3D printed catechin film based on the hydroxypropyl methylcellulose hydrogel bio-ink can further adjust the release of catechins according to the

Natural Polymers for Biomedical Applications, First Edition. Wenguo Cui and Lei Xiang.

adjustment of the ink concentration and the subsequent drying process [8]. Catechin has abundant hydroxyl groups and can be used to increase the adhesion of materials. It can self-assemble with keratin to form nanoparticles. When introduced into carboxymethylcellulose hydrogel, it can be used as a cross-linking point to enhance hydraulic coagulation. Glue viscosity can also meet the needs of hemostasis [9].

12.2 Catechins-Based Electrospinning for Biomedical Application

Electrospinning is an efficient method to produce nanofibers from all types of natural polymer solutions. The activity of catechins is susceptible to temperature, pH, oxygen, light, and many other conditions during processing. Therefore, electrospinning technology is expected to become an efficient technology for preparing catechin-stable materials. The combination of polysaccharides and polypeptides and other polymers (such as zein, pullulan, carboxymethyl cellulose, and Azivash gum) can effectively prepare electrospun fiber membranes loaded with catechins [10, 11]. Although the electrospun fiber with a single structure can maintain the stability of catechins as much as possible in the preparation stage, it cannot further control the release curve of catechins according to different microenvironments in the internal environment. By preparing electrospun fibers with a core–shell structure, the stability of catechins can be further ensured to avoid their burst release [12]. In addition, by controlling the concentration of PLGA and water-soluble catechin water-in-oil emulsions, it can be prepared according to actual needs. Electrospun fibers with different release phases can meet the diverse needs of biomedicine [13].

12.3 Catechins–Metal Complexes for Biomedical Application

The implantation of various artificial materials greatly increases the risk of infection at the lesion site. Among them, the formation of biofilm on the surface of the material is the main reason for the growth of bacteria and it is difficult to remove. At present, the commonly used method to prevent infection is to prepare an antibacterial coating on the surface of the material. Among them, silver nanoparticles are often used as the main functional component of broad-spectrum antibacterial. However, long-term exposure of silver-based materials in the body is likely to cause toxicity accumulation and bacterial resistance. Catechin has a variety of biologically active functions, including antibacterial ability, but it also has many defects such as stability and weak membrane permeability. Through the interaction of the polyhydroxyl group in catechins and metal ions, a metal polyphenol nano-network can be formed to improve its stability and antibacterial properties. For example, through the self-assembly of catechins and ions, a stable metal polyphenol nano-network can be formed. At the same time, catechins can provide rare earth ions with affinity for the skin. The rare earth ions can also enhance the membrane penetration ability of the overall material. Biofilm formation can play a good synergistic effect [14, 15]. In addition, experiments have shown that the combination of catechins and copper ions can significantly inhibit the proliferation of *Pseudomonas aeruginosa*, and the combination with ferrous ions can inhibit golden grapes activity of cocci [16]. In addition to the antibacterial ability, the inherent anti-tumor activity of catechins can also be further enhanced by the metal polyphenol nano-network. Nanoparticles formed by complexing with lanthanide metal ions can be rapidly internalized by cancer cells and degraded in an acidic environment at the endocrine stage to release lanthanide metal ions and catechins to promote the apoptosis of target cells and inhibit tumor cells the proliferation of [17]. In addition, nanoparticles composed of catechins and gold can be used to detect the lead content in urine samples. The lead–catechin complex formed on the surface of the nanoparticles can be expressed in the oxidation of AUR in the H_2O_2 medium [18].

References

1 Tang, G.Y., Zhao, C.N., Xu, X.Y. et al. (2019). Phytochemical composition and antioxidant capacity of 30 Chinese teas. *Antioxidants (Basel)* 8: 180.

2 Yang, C.S., Wang, X., Lu, G., and Picinich, S.C. (2009). Cancer prevention by tea: animal studies, molecular mechanisms and human relevance. *Nat. Rev. Cancer* 9: 429–439.

3 Zhao, C.N., Tang, G.Y., Cao, S.Y. et al. (2019). Phenolic profiles and antioxidant activities of 30 tea infusions from green, black, oolong, white, yellow and dark teas. *Antioxidants (Basel)* 8: 215.

4 Kim, A.R., Kim, K.M., Byun, M.R. et al. (2017). Catechins activate muscle stem cells by Myf5 induction and stimulate muscle regeneration. *Biochem. Biophys. Res. Commun.* 489: 142–148.

5 Lee, J.H., Jin, H.X., Shim, H.E. et al. (2010). Epigallocatechin-3-gallate inhibits osteoclastogenesis by down-regulating c-Fos expression and suppressing the nuclear factor-κB signal. *Mol. Pharmacol.* 77: 17–25.

6 Kim, E.S., Lee, J.S., and Lee, H.G. (2016). Calcium-alginate microparticles for sustained release of catechin prepared via an emulsion gelation technique. *Food Sci. Biotechnol.* 25: 1337–1343.

7 Lee, J.S., Kim, E.J., Chung, D., and Lee, H.G. (2009). Characteristics and antioxidant activity of catechin-loaded calcium pectinate gel beads prepared by internal gelation. *Colloids Surf. B* 74: 17–22.

8 Tagami, T., Yoshimura, N., Goto, E. et al. (2019). Fabrication of muco-adhesive oral films by the 3D printing of hydroxypropyl methylcellulose-based catechin-loaded formulations. *Biol. Pharm. Bull.* 42: 1898–1905.

9 Sun, Z., Chen, X.Y., Ma, X.M. et al. (2018). Cellulose/keratin-catechin nanocomposite hydrogel for wound hemostasis. *J. Mater. Chem. B* 6: 6133–6141.

10 Hoseyni, S.Z., Jafari, S.M., Tabarestani, H.S. et al. (2020). Production and characterization of catechin-loaded electrospun nanofibers from Azivash gum-polyvinyl alcohol. *Carbohydr. Polym.* 235: 115979.

11 Hoseyni, S.Z., Jafari, S.M., Tabarestani, H.S. et al. (2021). Release of catechin from Azivash gum-polyvinyl alcohol electrospun nanofibers in simulated food and digestion media. *Food Hydrocoll.* 112: 106366.

12 Arrieta, M.P., García, A.D., López, D. et al. (2019). Antioxidant bilayers based on PHBV and plasticized electrospun PLA-PHB fibers encapsulating catechin. *Nanomaterials (Basel)* 9: 346.

13 Ghitescu, R.E., Popa, A.M., Schipanski, A. et al. (2018). Catechin loaded PLGA submicron-sized fibers reduce levels of reactive oxygen species induced by MWCNT. *Eur. J. Pharm. Biopharm.* 122: 78–86.

14 Liu, L., Xiao, X., Li, K. et al. (2020). Prevention of bacterial colonization based on self-assembled metal-phenolic nanocoating from rare-earth ions and catechin. *ACS Appl. Mater. Interfaces* 12: 22237–22245.

15 Liu, L., Xiao, X., Li, K. et al. (2020). Synthesis of catechin-rare earth complex with efficient and broad-spectrum anti-biofilm activity. *Chem. Biodivers.* 17: e1900734.

16 Holloway, A.C., Mueller-Harvey, I., Gould, S.W.J. et al. (2012). The effect of copper(II), iron(II) sulphate, and vitamin C combinations on the weak antimicrobial activity of (+)-catechin against *Staphylococcus aureus* and other microbes. *Metallomics* 4: 1280–1286.

17 Li, K., Xiao, G., Richardson, J.J. et al. (2019). Targeted therapy against metastatic melanoma based on self-assembled metal-phenolic nanocomplexes comprised of green tea catechin. *Adv. Sci.* 6: 1801688.

18 Wu, Y.S., Huang, F.F., and Lin, Y.W. (2013). Fluorescent detection of lead in environmental water and urine samples using enzyme mimics of catechin-synthesized Au nanoparticles. *ACS Appl. Mater. Interfaces* 5: 1503–1509.

13

Sources, Structures, and Properties of Quercetin

Quercetin is a flavonol compound with multiple biological activities. It is mostly present in the stems, leaves, and fruits of plants in the form of glycosides (such as quercetin and hyperoside). Among them, the content of buckwheat stems and leaves is higher [1].

The phenolic hydroxyl structure of quercetin can directly scavenge free radicals in the body. At the same time, quercetin is also a natural metal ion chelator, which can chelate iron and copper ions in the body to reduce oxidative damage [2]. The antioxidant capacity of quercetin will be further strengthened when it is prepared into nanoparticles, and its solubility will also be improved [3]. Quercetin can also be involved in mediating the p53 signaling pathway to induce cell apoptosis [4–6], and its complex with metals will further improve its anti-tumor performance.

In addition, quercetin has the effects of lowering blood pressure, reducing capillary fragility, lowering blood lipids, and dilating coronary arteries. It also has an adjuvant therapeutic effect on patients with coronary heart disease and hypertension.

13.1 Quercetin-Based Hydrogel for Biomedical Application

Quercetin is a bioflavonoid with anti-inflammatory and antioxidant functions. It has been used as an oral supplemental treatment for osteoarthritis. Using hydrogel to encapsulate quercetin can further improve its bioavailability and achieve better therapeutic effects [7, 8]. Encapsulation of quercetin in PEG–PLGA hydrogels, protein hydrogels, or hydrogel microspheres prepared by ion cross-linking technology can make it stable under the acidic conditions of the stomach and released in the small intestine [9–11]. In addition, the multi-stage drug delivery system of hydrogel–liposome (or such as β-tricalcium phosphate nanoparticles)–quercetin is also an effective way to control the sustained release of quercetin [12]. By adjusting the components of the hydrogel carrier, its anti-tumor performance can be further enhanced. By using the property of hyaluronic acid to bind to the CD44 receptor, quercetin can be encapsulated in the hyaluronic acid hydrogel to realize the drug in the hyaluronic acid hydrogel [13].

Natural Polymers for Biomedical Applications, First Edition. Wenguo Cui and Lei Xiang.

13.2 Quercetin-Based Electrospinning for Biomedical Application

The electrospun fiber can effectively protect the stability of the drug loaded therein, and the structure of the electrospun fiber can be customized according to specific needs. Electrospinning with a core–shell structure is often used to protect the activity of loaded drugs and achieve sustained release. For example, quercetin is firstly compounded with chitosan to form nanoparticles and then used as a core material to be loaded on a shell of sodium alginate. The point-spun fiber can remain stable in acidic environments such as the stomach and be released directionally with the colon [14, 15]. In addition, the nanoparticle-embedded fiber, which has a similar effect to the shell–core structure electrospinning, can also be greatly improved. The coated drug reduces the sudden release phenomenon, and its inherent biological properties (such as sustained antibacterial properties in vitro) can be maintained for a long time [16–18]. Quercetin has the effect of regulating the oxidative stress state of tissues, and electrospun fiber membranes prepared based on this can significantly improve the neuropathy of diabetic rats induced by STZ [19]. On this basis, further bionic design of the fiber structure can provide richer physical and chemical clues for tissue repair. The quercetin-loaded 3D nanofibers composed of aligned and random nanofibers prepared by electrospinning technology have micro-scale pores in the outer layer, which can prevent the invasion of fibrous tissue while meeting the demand for material exchange, and the aligned fibrous structure inside. It is good for nerve repair and regeneration [20].

13.3 Quercetin–Metal Complexes for Biomedical Application

Quercetin is a well-known antioxidant and free radical scavenger. It can protect cells under pathological conditions caused by extracellular and intracellular oxidative stress caused by metal overload. It can chelate with free iron ions in cells to inhibit the generation of free radicals. In addition, quercetin can also act as a siderophore to promote the movement of iron between cells [21]. The nanocomplex formed by the complexation of quercetin and copper ions has the effect of promoting angiogenesis and calcium deposition, which can stimulate the expression of Runx2mRNA and protein, ALPmRNA, and type I collagen mRNA at the molecular level [22]. In addition, the hyaluronic acid–quercetin–copper complex can induce BRCA mutant triple-negative breast cancer poly-ADP-ribose polymerase inhibition and DNA damage through CD44 targeting under magnetic navigation, thereby minimizing the killing of normal human cells by tumor chemotherapy [23]. In addition, quercetin can significantly inhibit the overexpression of P-glycoprotein to reduce multidrug resistance in tumor treatment. Using gold nanocages as a carrier to deliver adriamycin and quercetin at the same time is a very important method. It is a promising nano-drug delivery system for overcoming multidrug-resistant tumors [24]. Nanoclusters prepared with ruthenium and quercetin can specifically detect the content

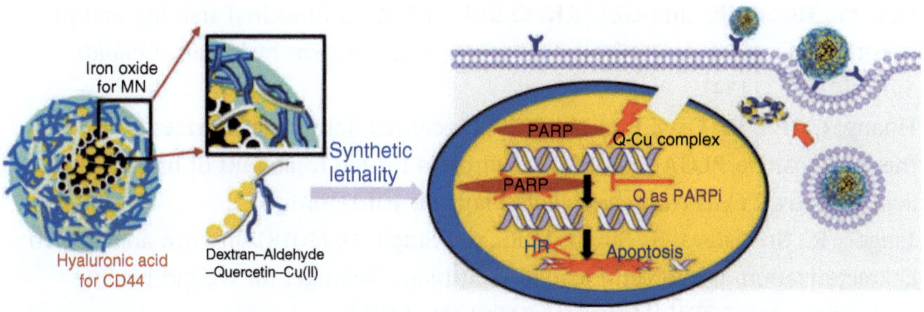

Figure 13.1 A novel nanomedicine based on natural materials for dual-targeted therapy of cancer. Source: Ref. [23]/Elsevier.

of heavy metal cobalt in cancer cells and can exhibit strong orange fluorescence in the cells under excitation at 465 nm. It is a combination of anti-cancer therapy and bio-imaging integrated multifunctional nanomaterials [25] (Figure 13.1).

References

1 David, A.V.A., Arulmoli, R., and Parasuraman, S. (2016). Overviews of biological importance of quercetin: a bioactive flavonoid. *Pharmacogn. Rev.* 10: 84–89.

2 Tang, Y.H., Li, Y.Y., Yu, H.Y. et al. (2014). Quercetin attenuates chronic ethanol hepatotoxicity: implication of "free" iron uptake and release. *Food Chem. Toxicol.* 67: 131–138.

3 Manca, M.L., Castangia, I., Caddeo, C. et al. (2014). Improvement of quercetin protective effect against oxidative stress skin damages by incorporation in nanovesicles. *Colloids Surf., B* 123: 566–574.

4 Suh, D.K., Lee, E.J., Kim, H.C., and Kim, J.H. (2010). Induction of G(1)/S phase arrest and apoptosis by quercetin in human osteosarcoma cells. *Arch. Pharm. Res.* 33: 781–785.

5 Tan, J., Wang, B.C., and Zhu, L.C. (2009). Regulation of survivin and Bcl-2 in HepG2 cell apoptosis induced by quercetin. *Chem. Biodivers.* 6: 1101–1110.

6 Zhang, Q., Zhao, X.H., and Wang, Z.J. (2008). Flavones and flavonols exert cytotoxic effects on a human oesophageal adenocarcinoma cell line (OE33) by causing G2/M arrest and inducing apoptosis. *Food Chem. Toxicol.* 46: 2042–2053.

7 Esposito, L., Barbosa, A.I., Moniz, T. et al. (2020). Design and characterization of sodium alginate and poly(vinyl) alcohol hydrogels for enhanced skin delivery of quercetin. *Pharmaceutics* 12: 1149.

8 Mok, S.W., Fu, S.C., Cheuk, Y.C. et al. (2020). Intra-articular delivery of quercetin using thermosensitive hydrogel attenuate cartilage degradation in an osteoarthritis rat model. *Cartilage* 11: 490–499.

9 Liu, K., Zha, X.Q., Shen, W.D. et al. (2020). The hydrogel of whey protein isolate coated by lotus root amylopectin enhance the stability and bioavailability of quercetin. *Carbohydr. Polym.* 236: 116009.

10 Dey, M., Ghosh, B., and Giri, T.K. (2020). Enhanced intestinal stability and pH sensitive release of quercetin in GIT through gellan gum hydrogels. *Colloids Surf., B* 196: 111341.

11 Huang, C., Fu, C., Qi, Z.P. et al. (2020). Localised delivery of quercetin by thermo-sensitive PLGA-PEG-PLGA hydrogels for the treatment of brachial plexus avulsion. *Artif. Cells Nanomed. Biotechnol.* 48: 1010–1021.

12 Jangde, R., Srivastava, S., Singh, M.R., and Singh, D. (2018). In vitro and in vivo characterization of quercetin loaded multiphase hydrogel for wound healing application. *Int. J. Biol. Macromol.* 115: 1211–1217.

13 Quagliariello, V., Armenia, E., Aurilio, C. et al. (2016). New treatment of medullary and papillary human thyroid cancer: biological effects of hyaluronic acid hydrogel loaded with quercetin alone or in combination to an inhibitor of aurora kinase. *J. Cell. Physiol.* 231: 1784–1795.

14 Wen, P., Zong, M.H., Hu, T.G. et al. (2018). Preparation and characterization of electrospun colon-specific delivery system for quercetin and its antiproliferative effect on cancer cells. *J. Agric. Food Chem.* 66: 11550–11559.

15 Wen, P., Hu, T.G., Li, L. et al. (2018). A colon-specific delivery system for quercetin with enhanced cancer prevention based on co-axial electrospinning. *Food Funct.* 9: 5999–6009.

16 Kost, B., Svyntkivska, M., Brzezinski, M. et al. (2020). PLA/β-CD-based fibres loaded with quercetin as potential antibacterial dressing materials. *Colloids Surf., B* 190: 110949.

17 Faraji, S., Nowroozi, N., Nouralishahi, A., and Shayeh, J.S. (2020). Electrospun poly-caprolactone/graphene oxide/quercetin nanofibrous scaffold for wound dressing: evaluation of biological and structural properties. *Life Sci.* 257: 118062.

18 Wang, Z., Zou, W., Liu, L.Y. et al. (2021). Characterization and bacteriostatic effects of β-cyclodextrin/quercetin inclusion compound nanofilms prepared by electrospinning. *Food Chem.* 338: 127980.

19 Thipkaew, C., Wattanathorn, J., and Muchimapura, S. (2017). Electrospun nanofibers loaded with quercetin promote the recovery of focal entrapment neuropathy in a rat model of streptozotocin-induced diabetes. *Biomed Res. Int.* 2017: 2017493.

20 Jang, S.R., Kim, J.I., Park, C.H., and Kim, C.S. (2020). The controlled design of electrospun PCL/silk/quercetin fibrous tubular scaffold using a modified wound coil collector and L-shaped ground design for neural repair. *Mater. Sci. Eng., C* 111: 110776.

21 Vlachodimitropoulou, E., Sharp, P.A., and Naftalin, R.J. (2011). Quercetin-iron chelates are transported via glucose transporters. *Free Radic. Biol. Med.* 50: 934–944.

22 Vimalraj, S., Rajalakshmi, S., Preeth, D.R. et al. (2018). Mixed-ligand copper(II) complex of quercetin regulate osteogenesis and angiogenesis. *Mater. Sci. Eng., C* 83: 187–194.

23 Cheng, H.W., Chiang, C.S., Ho, H.Y. et al. (2021). Dextran-modified quercetin-Cu(II)/hyaluronic acid nanomedicine with natural poly(ADP-ribose) polymerase inhibitor and dual targeting for programmed synthetic lethal therapy in triple-negative breast cancer. *J. Controlled Release* 329: 136–147.

24 Zhang, Z.P., Xu, S.H., Wang, Y. et al. (2018). Near-infrared triggered co-delivery of doxorubicin and quercetin by using gold nanocages with tetradecanol to maximize anti-tumor effects on MCF-7/ADR cells. *J. Colloid Interface Sci.* 509: 47–57.

25 Lakshmi, B.A., Bae, J.Y., An, J.H., and Kim, S. (2019). Nanoclusters prepared from ruthenium(II) and quercetin for fluorometric detection of cobalt(II), and a method for screening their anticancer drug activity. *Microchim. Acta* 186: 539.

14

Sources, Structures, and Properties of Resveratrol

Resveratrol is a kind of non-flavonoid polyphenol organic compound. Resveratrol has cis-conformation and trans-conformation. In 1940, resveratrol was first extracted from *Veratrum grandiflorum*, and it was also found in grape leaves. At present, resveratrol is mainly extracted from *Polygonum cuspidatum* and grape.

Resveratrol is an antitoxin secreted by plants, which is secreted in large quantities when plants are injured or infected, so it is called phytoalexin. Resveratrol has anti-inflammatory and antioxidant pharmacological effects [1]. After being ingested by the human body, resveratrol will be hydrolyzed quickly and then excreted through urine and feces. Its low bioavailability limits its better biological function. The researchers significantly improved its pharmacokinetic characteristics by modifying chemical groups and studying novel nano-drug delivery systems [2], and further expanded its applications in anticancer, antiviral, and cardiac protection [3–5]. Some studies have shown that resveratrol has broad-spectrum antibacterial activity and can inhibit the production of bacterial membrane [6] and based on these characteristics, resveratrol has great potential in the preparation of new antibacterial biomaterials.

14.1 Resveratrol-Based Hydrogel for Biomedical Application

Resveratrol is widely used in the treatment of various inflammatory diseases and malignant tumors due to its good anti-inflammatory and anti-tumor activity. Since the direct application of resveratrol itself has the problem of low bioavailability, designing various drug delivery systems has become a top priority. Using hydrogel as a drug carrier material can provide good environmental support for resveratrol to exert better pharmacological effects. Inflammation and insufficient angiogenesis are the main reasons leading to non-healing of wounds. The supramolecular host–guest gelatin hydrogel loaded with resveratrol and histone-1 can promote the expression of inflammatory factors interleukin 6, interleukin 1β, and tumor necrosis factor α increases transforming growth factor β1 and platelet endothelial cell adhesion. The expression of CD31 can inhibit inflammation of burn wounds and promote angiogenesis. This hydrogel has shear thinning properties that can meet

Natural Polymers for Biomedical Applications, First Edition. Wenguo Cui and Lei Xiang.

the needs of in situ injection to quickly adapt to irregular wounds [7]. In addition, excessive inflammation is considered to be the cause of scar formation. Resveratrol loaded into the peptide–hydrogel can inhibit the production of pro-inflammatory cytokines by macrophages, reduce inflammation, order collagen deposition, and ultimately reduce scar formation [8]. The main pathogenic factors of acne are common skin diseases caused by ductal occlusion caused by hyperkeratosis of the sebaceous glands and the secretion of androgens stimulated by the sebaceous glands. Applying resveratrol to the carboxymethyl cellulose gel mixed with resveratrol can significantly improve the occlusion of hair follicles and the symptoms of local purulent infection caused by acne [9]. In addition, compared with traditional creams, resveratrol ethanol body hydrogel can significantly improve the skin penetration parameters and skin deposition of resveratrol [10]. Resveratrol is used as the main anti-inflammatory component through chemical cross-linking with oxidized hyaluronic acid, which provides affinity for chondrocytes, to prepare oxidized hyaluronic acid/resveratrol hydrogel not only for growth factor carrier can also achieve the initial penetration of cells and subsequent integration with natural tissues. Oxidized hyaluronic acid/resveratrol hydrogel can be mixed with chondrocytes and simply injected into the degeneration or treatment site through a small needle to promote the gene expression of the main extracellular matrix components of chondrocytes-aggregin and type II collagen [11].

The intelligent response sodium deoxycholate hydrogel designed according to the pH value of the tumor extracellular microenvironment can simultaneously load doxorubicin and resveratrol and achieve sequential release. This delivery system can maintain proper DOX and resveratrol concentrations for a relatively long time, and reduces the exposure of systemic DOX in normal tissues. In addition, the synergy between DOX and resveratrol treatment resulted in a higher therapeutic effect on HeLa cell xenograft tumors in BALB/c nude mice [12]. In addition, ascites is a common complication of advanced liver cancer, and intraperitoneal chemotherapy is the current mainstream treatment. The use of temperature-sensitive hydrogels to deliver resveratrol and cisplatin can well solve the problem of rapid drug metabolism and the dilution of chemotherapeutic drugs in ascites, and achieve the induction of tumor cell apoptosis and promote cell cycle arrest at G1 [13].

In addition to directly cross-linking resveratrol with hydrogel structural components or directly embedding it in a hydrogel matrix, by designing multi-level drug delivery systems, such as liposome–hydrogel and nanoparticle–water a variety of delivery methods such as gel can achieve sustained release of resveratrol on a longer time scale, enhance its mucosal penetration and biological activity, and even give it a brain-targeting function by modifying the secondary carrier [14, 15].

14.2 Resveratrol-Based Electrospinning for Biomedical Application

Electrospinning technology provides a stable and reliable technical means for realizing the local controllable release of resveratrol. By preparing resveratrol nanocrystals, the problem of low water solubility in the application process can be significantly solved. The physical adsorption method can integrate resveratrol nanocrystals and polyester fiber nanofibers, and can achieve strong antibacterial activity against infections on the skin surface [16]. Coaxial electrospun fibers prepared based on polycaprolactone and gelatin can effectively control the release of resveratrol and improve its bioavailability. The results of controlled release of resveratrol for five days show that resveratrol can reach the required therapeutic concentration locally and induce a higher apoptotic effect [17]. The electrospun fiber has high encapsulation efficiency as a drug carrier, and the multi-layer fiber components can be individually selected according to the characteristics of the microenvironment of the target site. For example, oral enteral administration needs to overcome the problem of drugs being decomposed and inactivated in the acidic environment of the stomach. By loading resveratrol in nano-electrospun fibers composed mainly of chitosan and gelatin, the release of drugs under acidic conditions can be inhibited to ensure that more active ingredients of the drug can function in the intestine [18], a nanofiber drug delivery system loaded with resveratrol prepared based on polylactic acid can achieve accelerated release in an acidic environment, and can play an antibacterial effect in an acidic microenvironment such as when bacterial infections occur in the oral cavity, while in a neutral environment it acts as a drug reservoir to preserve drug activity [19]. Resveratrol electrospinning drug delivery system has practical value in reducing the development of drug resistance. Co-administration of multiple drugs is a common strategy to overcome drug resistance. Resveratrol and yellow can be achieved by electrospinning with a putamen structure. Humicol has a coordinated and sustained release for up to 350 hours, and the combination of the two can significantly inhibit the proliferation of breast cancer cells and largely avoid the development of drug resistance [20]. In addition to resveratrol's widely used anti-inflammatory, antibacterial, and anti-tumor activities, its more biologically active functions have been discovered by researchers. For example, the polylactic acid–gelatin porous electrospinning scaffold loaded with resveratrol can promote cartilage repair by regulating the PI3K/AKT signaling pathway [21], and the polycaprolactone electrospinning tubular structure loaded with resveratrol can induce a large amount of M2 macrophages penetrate the graft wall, and at the same time increase the production of NO and the ability to form tubes to achieve rapid endothelialization of transplanted small blood vessels [22].

14.3 Resveratrol–Metal Complexes for Biomedical Application

Resveratrol is a natural reducing agent that reduces such things as gold and silver ion chemistry to gold nanoparticles and silver nanoparticles without the need for additional reducing agents, in which hydroxyl and C=C on resveratrol aromatic rings are involved in the reduction reaction. Resveratrol–gold (silver) metal complexes showed stronger antibacterial activity against Gram-negative and -positive bacteria compared to their monomers [23]. Copper and zinc ions are the main metal ions in the nucleus, the concentration of copper in serum and tissue is significantly increased in various malignant tumors, as a natural antioxidant, resveratrol can cause DNA degradation of cells such as lymphocytes in the presence of transition metal ions such as copper, which may be one of the mechanisms of resveratrol with high anti-cancer activity, based on this design resveratrol–metal anti-cancer complex is also widely explored its application value [24]. In addition, drug delivery studies designed by resveratrol–gold nanocomposites showed that the drug effect of metal complex systems increased by 65% compared to resveratrol alone. In vitro drug release observations show that drug release of resveratrol–gold complex is pH-responsive, maintains a high degree of stability and biocompatibility in physiological conditions (pH 7.4), and releases 95% of drugs under acidic conditions (pH 5.2), which also demonstrates its potential for application in intelligent drug delivery systems [25]. Further development of the application of resveratrol metal complexes in the field of cancer treatment is the addition of tumor imaging function, by marking the appropriate radionuclides, resveratrol tumor targeting can be used radionuclide imaging technology to locate tumor sites in the body. The 99mTc-labeled resveratrol–gold nanocomposites show better in vivo intake and retention and significantly improve their positioning and intake at the tumor site for use in cancer imaging [26].

References

1 Pannu, N. and Bhatnagar, A. (2019). Resveratrol: from enhanced biosynthesis and bioavailability to multitargeting chronic diseases. *Biomed. Pharmacother.* 109: 2237–2251.

2 Chimento, A., De Amicis, F., Sirianni, R. et al. (2019). Progress to improve oral bioavailability and beneficial effects of resveratrol. *Int. J. Mol. Sci.* 20: 1381.

3 Meng, X., Zhou, J., Zhao, C.N. et al. (2020). Health benefits and molecular mechanisms of resveratrol: a narrative review. *Foods* 9: 340.

4 Yu, X.D., Yang, J.L., Zhang, W.L., and Liu, D.X. (2016). Resveratrol inhibits oral squamous cell carcinoma through induction of apoptosis and G2/M phase cell cycle arrest. *Tumour Biol.* 37: 2871–2877.

5 Militaru, C., Donoiu, I., Craciun, A. et al. (2013). Oral resveratrol and calcium fructoborate supplementation in subjects with stable angina pectoris: effects on lipid profiles, inflammation markers, and quality of life. *Nutrition* 29: 178–183.

6 Vestergaard, M. and Ingmer, H. (2019). Antibacterial and antifungal properties of resveratrol. *Int. J. Antimicrob. Agents* 53: 716–723.

7 Zheng, Y.Y., Yuan, W.H., Liu, H.L. et al. (2020). Injectable supramolecular gelatin hydrogel loading of resveratrol and histatin-1 for burn wound therapy. *Biomater. Sci.* 8: 4810–4820.

8 Zhao, C.C., Zhu, L., Wu, Z. et al. (2020). Resveratrol-loaded peptide-hydrogels inhibit scar formation in wound healing through suppressing inflammation. *Regener. Biomater.* 7: 99–108.

9 Fabbrocini, G., Staibano, S., De Rosa, G. et al. (2011). Resveratrol-containing gel for the treatment of acne vulgaris: a single-blind, vehicle-controlled, pilot study. *Am. J. Clin. Dermatol.* 12: 133–141.

10 Arora, D. and Nanda, S. (2019). Quality by design driven development of resveratrol loaded ethosomal hydrogel for improved dermatological benefits via enhanced skin permeation and retention. *Int. J. Pharm.* 567: 118448.

11 Sheu, S.Y., Chen, W.S., Sun, J.S. et al. (2013). Biological characterization of oxidized hyaluronic acid/resveratrol hydrogel for cartilage tissue engineering. *J. Biomed. Mater. Res. A* 101: 3457–3466.

12 Mekonnen, T.W., Andrgie, A.T., Darge, H.F. et al. (2020). Bioinspired composite, pH-responsive sodium deoxycholate hydrogel and generation 4.5 Poly(amidoamine) dendrimer improves cancer treatment efficacy via doxorubicin and resveratrol co-delivery. *Pharmaceutics* 12: 1069.

13 Wen, Q., Zhang, Y., Luo, J. et al. (2020). Therapeutic efficacy of thermosensitive pluronic hydrogel for codelivery of resveratrol microspheres and cisplatin in the treatment of liver cancer ascites. *Int. J. Pharm.* 582: 119334.

14 Rajput, A., Bariya, A., Allam, A. et al. (2018). In situ nanostructured hydrogel of resveratrol for brain targeting: in vitro–in vivo characterization. *Drug Deliv. Transl. Res.* 8: 1460–1470.

15 Marycz, K., Smieszek, A., Trynda, J. et al. (2019). Nanocrystalline hydroxyapatite loaded with resveratrol in colloidal suspension improves viability, metabolic activity and mitochondrial potential in human adipose-derived mesenchymal stromal stem cells (hASCs). *Polymers (Basel)* 11: 92.

16 Karakucuk, A. and Tort, S. (2020). Preparation, characterization and antimicrobial activity evaluation of electrospun PCL nanofiber composites of resveratrol nanocrystals. *Pharm. Dev. Technol.* 25: 1216–1225.

17 Al-Attar, T. and Madihally, S.V. (2018). Influence of controlled release of resveratrol from electrospun fibers in combination with siRNA on leukemia cells. *Eur. J. Pharm. Sci.* 123: 173–183.

18 Rostami, M., Ghorbani, M., Mohammadi, M.A. et al. (2019). Development of resveratrol loaded chitosan-gellan nanofiber as a novel gastrointestinal delivery system. *Int. J. Biol. Macromol.* 135: 698–705.

19 Bonadies, I., Di Cristo, F., Valentino, A. et al. (2020). pH-responsive resveratrol-loaded electrospun membranes for the prevention of implant-associated infections. *Nanomaterials (Basel)* 10: 1175.

20 Zhang, X., Han, L.B., Sun, Q.H. et al. (2020). Controlled release of resveratrol and xanthohumol via coaxial electrospinning fibers. *J. Biomater. Sci. Polym. Ed.* 31: 456–471.

21 Yu, F., Li, M., Yuan, Z.P. et al. (2018). Mechanism research on a bioactive resveratrol-PLA-gelatin porous nano-scaffold in promoting the repair of cartilage defect. *Int. J. Nanomed.* 13: 7845–7858.

22 Wang, Z.H., Wu, Y.F., Wang, J.N. et al. (2017). Effect of resveratrol on modulation of endothelial cells and macrophages for rapid vascular regeneration from electrospun poly(ε-caprolactone) scaffolds. *ACS Appl. Mater. Interfaces* 9: 19541–19551.

23 Park, S., Cha, S.H., Cho, I. et al. (2016). Antibacterial nanocarriers of resveratrol with gold and silver nanoparticles. *Mater. Sci. Eng., C* 58: 1160–1169.

24 Azmi, A.S., Bhat, S.H., and Hadi, S.M. (2005). Resveratrol-Cu(II) induced DNA breakage in human peripheral lymphocytes: implications for anticancer properties. *FEBS Lett.* 579: 3131–3135.

25 Kumar, C.G., Poornachandra, Y., and Mamidyala, S.K. (2014). Green synthesis of bacterial gold nanoparticles conjugated to resveratrol as delivery vehicles. *Colloids Surf., B* 123: 311–317.

26 Kamal, R., Chadha, V.D., and Dhawan, D.K. (2018). Physiological uptake and retention of radiolabeled resveratrol loaded gold nanoparticles (Tc-Res-AuNP) in colon cancer tissue. *Nanomedicine* 14: 1059–1071.

15

Sources, Structures, and Properties of Curcumin

Curcumin is a diketone compound extracted from the rhizomes of Zingiberaceae and Araceae plants. Curcumin was first discovered by Vogel and Pelletier in 1815, and was obtained as a pure compound in 1842. The bis-α, β-unsaturated β-diketone with keto-enol tautomerism chemical structure of curcumin was first reported in 1910.

Subsequent research on its physiological and pharmacological effects has made significant progress. Its effects in anti-inflammatory, anti-oxidation, scavenging of oxygen free radicals, anti-infection, anti-fibrosis, and anti-cancer have been revealed, which may be related to its inhibition of nuclear factor-κB and activin-1 that are related to the activation and expression of transcription factors [1]. In addition, studies have found that eating foods rich in curcumin every day can reduce the incidence of Alzheimer's disease, and experiments have shown that curcumin-induced exosomes inhibit the hyperphosphorylation of Tau protein through the AKT/GSK-3β pathway, thereby effectively improving the cognitive function of Alzheimer's disease mice [2]. Researchers use nanotechnology to improve its low water solubility and combine it with targeted drug delivery to further enrich the application of curcumin in the biomedical field [3–7].

15.1 Curcumin-Based Hydrogel for Biomedical Application

Curcumin is a polyphenol that is more effective than many chemotherapy drugs. It can inhibit multiple signaling pathways at the same time, leading to the regulation and down-regulation of multiple oncogenic activities, tumor suppressor genes, multiple transcription factors, and their signaling pathways. Loading curcumin through various hydrogels is currently a common method to improve its utilization. Nanocellulose–chitosan hydrogels prepared by chemical cross-linking can provide a good carrier platform for curcumin and improve curcumin's performance. The solubility of the drug in turn increases its absorption in the stomach and intestinal epithelium [8].

Natural Polymers for Biomedical Applications, First Edition. Wenguo Cui and Lei Xiang.

In addition, the curcumin sustained-release system based on lotus root pullulan–chitosan composite hydrogel can improve the solubility and stability of the drug while ensuring the stability of the drug in the gastric acid environment [9]. Chemically cross-linked hydrogels loaded with curcumin always have problems such as potential biological toxicity and the inability to achieve long-term sustained release of drugs. Curcumin itself has the ability to couple with functional groups located on the polymer chain. Cross-linking through biodegradable carbonate bonds, using curcumin as part of the polymer backbone, can protect the activity of curcumin and further improve its mucosa. Permeability can achieve sustained release for up to 80 days [10, 11].

The combination with cyclodextrins can also effectively improve the solubility and stability of curcumin, and the hydrogel film prepared based on this can achieve sustained drug release, which is a potential choice for wound dressings [12, 13]. Based on the hydrophobic properties of curcumin and the engineering self-assembly of peptides, it is possible to realize the encapsulation of curcumin in the hydrogel network and the self-assembly of peptides at the same time, and the addition of curcumin makes it reside in the hydrophobic core of the fiber to bridge the difference. The fibers thereby create an additional cross-linked network in the hydrogel network to further strengthen the mechanical properties of the hydrogel. At the same time, curcumin released from the hydrogel can effectively induce caspase3-mediated programmed cell death, thus verifying the effective protection of its biological activity [14]. The hydrophobic effect of curcumin combined with the non-polar region of PEGDA can increase the encapsulation efficiency of the drug to nearly 100% without the need for other chemical cross-linking methods. By adding microporous starch, the lysozyme in the microenvironment can be effectively adsorbed to achieve the purpose of high concentration of curcumin at the tumor site. In addition, curcumin can also be used as a chemotherapeutic drug sensitizer. The pH is synthesized by free radical polymerization. Temperature-responsive hydrogel can achieve high efficiency and controllable release of curcumin and doxorubicin, enhance the therapeutic effect of chemotherapeutic drugs, and help avoid the development of drug resistance [15, 16].

Regardless of tumor treatment or tissue repair, the overall treatment process is the result of multiple factors, which rely on the synergy of multiple signaling molecules. Take wound healing as an example. In the process of wound healing, it is necessary to promote the mitosis of wound cells and protect the whole tissue from oxidative damage by designing a new dual-drug co-loaded in situ gel nanoparticle/hydrogel system. This facilitates the synergistic release of epidermal growth factor and curcumin, leading to more granulation tissue formation, collagen deposition, and capillary regeneration in the established full-thickness excision wound model [17]. Utilizing the hydrophobic properties of curcumin, and encapsulating it in polymer micelles is also an effective means to improve its encapsulation efficiency. On the basis of this encapsulation strategy, the construction of self-healing hydrogel based on dynamic Schiff base can be very good. It meets the problem that the skin of the joints is frequently stretched and the excipients fail [18, 19] (Figure 15.1).

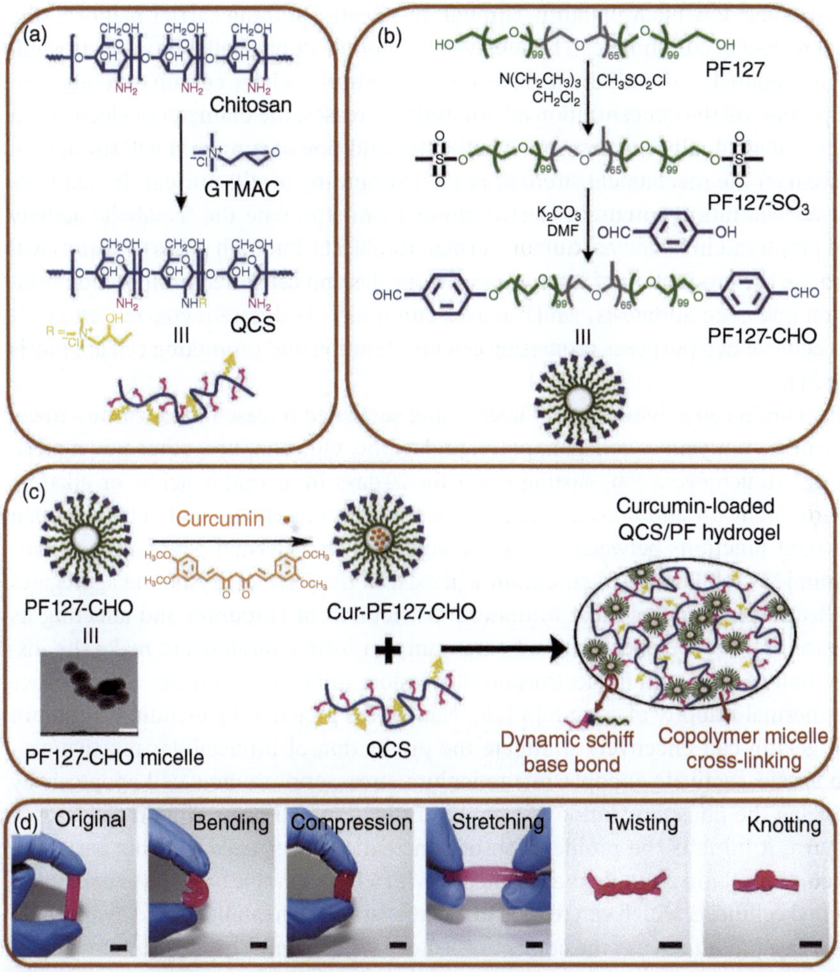

Figure 15.1 (a) Synthesis route of QCS polymer. (b) Structure of PF127-CHO polymer. (c) Schematic of Cur-QCS/PF hydrogel and TEM images of PF127-CHO micelles. (d) Rhodamine B stained images of QCS/PF1.0 hydrogel in various deformed states. Source: Ref. [19]/Elsevier.

15.2 Curcumin-Based Electrospinning for Biomedical Application

Electrospinning represents a new generation of medical textiles and has broad application prospects in soft-tissue repair and local drug delivery. By combining with biologically active molecules, electrospun fibers can obtain different characteristics. At the same time, the diverse structures of electrospun fibers also provide a good carrier platform for the human body to use various biologically active substances.

As a natural polyphenol, curcumin has antioxidant, anti-inflammatory, antibacterial, and anti-cancer activities, but its insolubility and circulatory elimination result in extremely low bioavailability. Through the electrospinning carrier platform, the local release of curcumin can be achieved to improve its bioavailability. Electrospun fibers prepared under different curcumin concentrations have certain differences in properties. As the concentration of curcumin increases, the diameter of electrospun fibers will gradually decrease, but whether the addition of curcumin will strengthen or weaken the mechanical strength of electrospinning is still unclear. In addition, low-concentration curcumin electrospinning can stimulate the metabolic activity and proliferation of normal human dermal fibroblasts, but high doses (>10 μm) will increase the production of reactive oxygen species and cause cell proliferation inhibition and even apoptosis, which makes curcumin. At different concentrations, it can achieve two purposes promoting cell proliferation and promoting cell apoptosis [20, 21].

Curcumin can achieve high efficiency and sustained release in electrospun fibers prepared from yellow collagen, polycaprolactone, chitosan, and other raw materials. It can achieve a long-lasting effect for 20 days in a weakly acidic or alkaline environment. In addition, the interaction between nanofibers and cells can open the tight junctions between cells to enhance the transdermal penetration of curcumin [22–24]. Untreated curcumin will exist in the form of crystalline aggregates in electrospinning, resulting in uneven distribution of curcumin and affecting its release profile. Cyclodextrins and curcumin can form complexes to make the distribution of curcumin in electrospun fibers more uniform, and it can also improve the thermal stability of curcumin [25]. Nanofibers prepared by blending curcumin and gelatin can effectively stimulate the production of intracellular reactive oxygen species, activate endoplasmic reticulum stress, and promote cell apoptosis by reducing the phosphorylation of signal transduction and transcription activator 3. In turn, it inhibits the proliferation of pancreatic cancer cells. On this basis, the curcumin–gelatin nanofibers are wrapped with a hydrophobic outer layer composed of ethyl cellulose, which can reduce the water vapor permeability of the hydrophilic inner layer and increase the contact resistance with water. The close interaction of hydrogen bonds between them improves thermal stability [21, 26].

15.3 Curcumin–Metal Complexes for Biomedical Application

The formation of complexes between curcumin and metals can improve the poor water solubility and low bioavailability of curcumin itself. Compared with curcumin alone, it may show greater biological activity. In addition, metals have also become important therapeutic agents in the medical field in recent years, showing the potential to improve drug resistance and toxicity. The role of curcumin–metal complexes in anti-cancer therapy has been widely reported.

The preparation of curcumin–iron and curcumin–boron complexes can improve its bioavailability while maintaining the same anti-cancer properties as the parent compound. Curcumin–iron metal complexes can increase the relative expression of the two key pro-apoptotic proteins: cytochrome c and cleaved caspase-3, and the anti-apoptotic protein heme oxygenase-1 [27]. The complex of curcumin and palladium can induce the growth inhibition and apoptosis of human prostate cancer cells through the production of reactive oxygen species and JNK phosphorylation related to the down-regulation of GSTP1 [28].

In addition, curcumin–metal complexes can also be combined with the current hot photodynamic anti-cancer treatment strategies. Curcumin and platinum, zinc, nickel, and other metal complexes have photoactivated cytotoxicity, which can generate reactive oxygen species under the excitation of photoelectrons. At the same time, the mitochondrial membrane potential is reduced under light irradiation, thereby blocking the cell cycle process in the sub-G1 phase and inducing cell apoptosis [29–31].

Curcumin itself has a certain degree of reducibility. At present, there have been reports that curcumin itself can be used to reduce gold ions directly in the water phase to form a functionalized curcumin–gold metal complex and exhibit strong antioxidant activity. In addition, the presence of curcumin makes gold sol stable for several months. Curcumin and manganese complexes can enter the brain and significantly improve ischemia–reperfusion injury by preventing brain lipid peroxidation [32, 33]. In addition to antioxidant effects, gold nanoparticles can inhibit the formation of osteoclasts and have great potential in preventing osteoporosis. The complexes formed with curcumin can inhibit the destruction of bone marrow stromal cells by inhibiting RANKL-induced signaling pathways and prevent osteocyte differentiation [34].

References

1 Edwards, R.L., Luis, P.B., Varuzza, P.V. et al. (2017). The anti-inflammatory activity of curcumin is mediated by its oxidative metabolites. *J. Biol. Chem.* 292: 21243–21252.

2 Wang, H., Sui, H.J., Zheng, Y. et al. (2019). Curcumin-primed exosomes potently ameliorate cognitive function in AD mice by inhibiting hyperphosphorylation of the Tau protein through the AKT/GSK-3β pathway. *Nanoscale* 11: 7481–7496.

3 Zheng, Z.G., Zhang, X.C., Carbo, D. et al. (2010). Sonication-assisted synthesis of polyelectrolyte-coated curcumin nanoparticles. *Langmuir* 26: 7679–7681.

4 Lvov, Y.M., Pattekari, P., Zhang, X.C., and Torchilin, V. (2011). Converting poorly soluble materials into stable aqueous nanocolloids. *Langmuir* 27: 1212–1217.

5 Chen, X., Chen, Y., Zou, L.Q. et al. (2019). Plant-based nanoparticles prepared from proteins and phospholipids consisting of a core multilayer-shell structure: fabrication, stability, and foamability. *J. Agric. Food Chem.* 67: 6574–6584.

6 Wang, Y., Gao, D., Liu, Y. et al. (2021). Immunogenic-cell-killing and immunosuppression-inhibiting nanomedicine. *Bioact. Mater.* 6: 1513–1527.

7 Yang, J.L., Zhang, X.C., Liu, C. et al. (2021). Biologically modified nanoparticles as theranostic bionanomaterials. *Prog. Mater. Sci.* 118: 100768.

8 Gunathilake, T.M.S.U., Ching, Y.C., Chuah, C.H. et al. (2018). Influence of a nonionic surfactant on curcumin delivery of nanocellulose reinforced chitosan hydrogel. *Int. J. Biol. Macromol.* 118: 1055–1064.

9 Liu, K., Huang, R.L., Zha, X.Q. et al. (2020). Encapsulation and sustained release of curcumin by a composite hydrogel of lotus root amylopectin and chitosan. *Carbohydr. Polym.* 232: 115810.

10 Hanafy, N.A.N., Leporatti, S., and El-Kemary, M. (2020). Mucoadhesive curcumin crosslinked carboxy methyl cellulose might increase inhibitory efficiency for liver cancer treatment. *Mater. Sci. Eng. C* 116: 111119.

11 Shpaisman, N., Sheihet, L., Bushman, J. et al. (2012). One-step synthesis of biodegradable curcumin-derived hydrogels as potential soft tissue fillers after breast cancer surgery. *Biomacromolecules* 13: 2279–2286.

12 Wathoni, N., Motoyama, K., Higashi, T. et al. (2017). Enhancement of curcumin wound healing ability by complexation with 2-hydroxypropyl-γ-cyclodextrin in sacran hydrogel film. *Int. J. Biol. Macromol.* 98: 268–276.

13 Kiti, K. and Suwantong, O. (2020). Bilayer wound dressing based on sodium alginate incorporated with curcumin-β-cyclodextrin inclusion complex/chitosan hydrogel. *Int. J. Biol. Macromol.* 164: 4113–4124.

14 Altunbas, A., Lee, S.J., Rajasekaran, S.A. et al. (2011). Encapsulation of curcumin in self-assembling peptide hydrogels as injectable drug delivery vehicles. *Biomaterials* 32: 5906–5914.

15 Ning, P.A., Lü, S.Y., Bai, X. et al. (2018). High encapsulation and localized delivery of curcumin from an injectable hydrogel. *Mater. Sci. Eng., C* 83: 121–129.

16 Abedi, F., Davaran, S., Hekmati, M. et al. (2021). An improved method in fabrication of smart dual-responsive nanogels for controlled release of doxorubicin and curcumin in HT-29 colon cancer cells. *J. Nanobiotechnol.* 19: 18.

17 Li, X.L., Ye, X.L., Qi, J.Y. et al. (2016). EGF and curcumin co-encapsulated nanoparticle/hydrogel system as potent skin regeneration agent. *Int. J. Nanomed.* 11: 3993–4009.

18 Gong, C.Y., Wu, Q.J., Wang, Y.J. et al. (2013). A biodegradable hydrogel system containing curcumin encapsulated in micelles for cutaneous wound healing. *Biomaterials* 34: 6377–6387.

19 Qu, J., Zhao, X., Liang, Y.P. et al. (2018). Antibacterial adhesive injectable hydrogels with rapid self-healing, extensibility and compressibility as wound dressing for joints skin wound healing. *Biomaterials* 183: 185–199.

20 Tsekova, P.B., Spasova, M.G., Manolova, N.E. et al. (2017). Electrospun curcumin-loaded cellulose acetate/polyvinylpyrrolidone fibrous materials with complex architecture and antibacterial activity. *Mater. Sci. Eng., C* 73: 206–214.

21 Mouthuy, P.A., Skoc, M.S., Gasparovic, A.C. et al. (2017). Investigating the use of curcumin-loaded electrospun filaments for soft tissue repair applications. *Int. J. Nanomed.* 12: 3977–3991.

22 Faralli, A., Shekarforoush, E., Ajalloueian, F. et al. (2019). permeability enhancement of curcumin across Caco-2 cells monolayers using electrospun xanthan-chitosan nanofibers. *Carbohydr. Polym.* 206: 38–47.

23 Guo, G., Fu, S.Z., Zhou, L.X. et al. (2011). Preparation of curcumin loaded poly(ε-caprolactone)-poly(ethylene glycol)-poly(ε-caprolactone) nanofibers and their in vitro antitumor activity against Glioma 9L cells. *Nanoscale* 3: 3825–3832.

24 Ranjbar-Mohammadi, M. and Bahrami, S.H. (2016). Electrospun curcumin loaded poly(ε-caprolactone)/gum tragacanth nanofibers for biomedical application. *Int. J. Biol. Macromol.* 84: 448–456.

25 Sun, X.Z., Williams, G.R., Hou, X.X., and Zhu, L.M. (2013). Electrospun curcumin-loaded fibers with potential biomedical applications. *Carbohydr. Polym.* 94: 147–153.

26 Cheng, T., Zhang, Z.H., Shen, H. et al. (2020). Topically applicated curcumin/gelatin-blended nanofibrous mat inhibits pancreatic adenocarcinoma by increasing ROS production and endoplasmic reticulum stress mediated apoptosis. *J. Nanobiotechnol.* 18: 126.

27 Mohammed, F., Rashid-Doubell, F., Taha, S. et al. (2020). Effects of curcumin complexes on MDA-MB-231 breast cancer cell proliferation. *Int. J. Oncol.* 57: 445–455.

28 Valentini, A., Conforti, F., Crispini, A. et al. (2009). Synthesis, oxidant properties, and antitumoral effects of a heteroleptic palladium(II) complex of curcumin on human prostate cancer cells. *J. Med. Chem.* 52: 484–491.

29 Upadhyay, A., Gautam, S., Ramu, V. et al. (2019). Photocytotoxic cancer cell-targeting platinum(II) complexes of glucose-appended curcumin and biotinylated 1,10-phenanthroline. *Dalton Trans.* 48: 17556–17565.

30 Qin, L.Q., Liang, C.J., Zhou, Z. et al. (2021). Mitochondria-localizing curcumin-cryptolepine Zn(II) complexes and their antitumor activity. *Bioorg. Med. Chem.* 30: 115948.

31 Banaspati, A., Raza, M.K., and Goswami,.T.K. (2020). Ni(II) curcumin complexes for cellular imaging and photo-triggered *in vitro* anticancer activity. *Eur. J. Med. Chem.* 204: 112632.

32 Singh, D.K., Jagannathan, R., Khandelwal, P. et al. (2013). *In situ* synthesis and surface functionalization of gold nanoparticles with curcumin and their antioxidant properties: an experimental and density functional theory investigation. *Nanoscale* 5: 1882–1893.

33 Vajragupta, O., Boonchoong, P., Watanabe, H. et al. (2003). Manganese complexes of curcumin and its derivatives: evaluation for the radical scavenging ability and neuroprotective activity. *Free Radic. Biol. Med.* 35: 1632–1644.

34 Heo, D.N., Ko, W.K., Moon, H.J. et al. (2014). Inhibition of osteoclast differentiation by gold nanoparticles functionalized with cyclodextrin curcumin complexes. *ACS Nano* 8: 12049–12062.

16
Summary

Polyphenols are secondary metabolites of plants, which have at least one hydroxyl aromatic ring. Polyphenols are natural polymers widely found in natural plants [1–3]. According to their chemical structure, polyphenols can be divided into the following categories: phenolic acids (lignans), flavonoids (catechins, quercetin, kaempferol, anthocyanins), stilbene (resveratrol), and polyphenolic amides (avenanthramides and capsaicin). Phenolic compounds generally have the ability of antioxidation and scavenging free radicals, which can effectively prevent lipid peroxidation and cell oxidative damage mediated by harmful free radicals. This property is related to the ability of phenolic compounds to provide hydrogen atoms or chelate metal cations. In recent years, the combination of polyphenols and nanotechnology (such as polyphenol nanoparticles, metal–polyphenol nanonetworks) has further expanded their applications in antioxidation, anticancer, antiviral, and cardiovascular protection.

References

1 Tam, R.Y., Smith, L.J., and Shoichet, M.S. (2017). Engineering cellular microenvironments with photo- and enzymatically responsive hydrogels: toward biomimetic 3D cell culture models. *Acc. Chem. Res.* 50: 703–713.
2 Zhang, Q.J., Li, W., Li, K. et al. (2020). The chromosome-level reference genome of tea tree unveils recent bursts of non-autonomous LTR retrotransposons in driving genome size evolution. *Mol. Plant* 13: 935–938.
3 Parekh, G., Shi, Y., Zheng, J. et al. (2018). Nano-carriers for targeted delivery and biomedical imaging enhancement. *Ther. Delivery* 9: 451–468.

Natural Polymers for Biomedical Applications, First Edition. Wenguo Cui and Lei Xiang.

17

Conclusion and Outlook

In this book, we divide the natural polymer materials that are widely used in the field of biomedicine into polysaccharides, peptides, and polyphenols, and briefly summarize their sources, structures, and properties. We focused on the application of natural polymers in the biomedical field in the form of electrospun fibers, 3D printed scaffolds metal complexes, and hydrogels [1–12]. The ideas and starting points of natural polymer biomaterial design can be briefly divided into the following four categories:

(1) Microenvironment responsiveness, that is, the material itself can be such as ROS, inflammatory factors, acidity, MMP-13, and other biological microenvironment of chemical signals or pressure, friction, and other physical signals to stimulate the material's intelligent response changes [13–15]. For example, bio-orthogonal patch design, acidity, gelatin can be MMP-13 responsive degradation. In addition, micro-therapeutic robots powered by microintrinsic changes are also hot areas of research.

(2) Target delivery system: This category can be divided into signal-mediated target system, spatial positioning target system, and external navigation target delivery system [16, 17]. Signal-mediated targeting systems use specific structures in natural polymers to combine with cell protein receptors to achieve material–cell target delivery (e.g. hyaluronic acids can be combined with CD44 receptor specificity). This material can sense changes in the surrounding magnetic field in the body and make directional movements depending on the direction of the magnetic field.

(3) Bionic structure type [2, 18]: This type is inspired by the physiological structure of natural plants and animals or other mechanical field structures and the use of natural polymers to achieve a breakthrough in the properties of materials (e.g. bionic lotus structure) or to achieve the application scenario of clever transplantation (e.g. bearing-inspired joint cavity lubrication microspheres).

(4) External-stimulation-mediated type: This kind of difference with the first category mainly lies in the external stimulus signal that is not applied. External-stimulation-mediated materials are widely used in the magnetic heat or photothermal treatment of tumors [19, 20].

Natural Polymers for Biomedical Applications, First Edition. Wenguo Cui and Lei Xiang.

However, the current research on natural polymers is far from sufficient, and with the advent of the era of smart medicine and precision medicine, new challenges have been proposed for the design and preparation of biomedical materials, and new opportunities have also been brought about. We believe that natural polymers still need further research and exploration in the following aspects.

(1) Explore the difference in properties between natural polymers from different sources, and formulate relevant industry standards to reduce the batch-to-batch quality differences of natural polymers.
(2) Natural polymers have different molecular weights, and there is still a lack of the influence of the molecular weight of natural polymers on their related physical and chemical properties.
(3) Polypeptide natural polymers such as collagen and fibrin are often used to simulate the extracellular matrix environment. Polysaccharides also play an important role in signal transmission between cells. However, the current special sequence and the relationship between cell fates still need to be further clarified and enriched, and suitable natural polymer materials are selected to achieve cell function regulation, cell-directed differentiation, and local recruitment of specific cells.
(4) Utilizing the conductivity of natural polymers to explore its application in artificial skin, biosensors, etc. has great potential.
(5) From the perspective of bionics and natural phenomena, it is a big challenge to develop and design natural polymer materials that can fully simulate the physiological structure environment of target tissues from micro to macro.
(6) It is a challenge work to prepare natural polymer tissue engineering materials with large size (meeting clinical needs) and satisfying mechanical properties and good biological activity functions.
(7) Identify specific domains that work in natural polymers through machine-learning-based artificial intelligence.
(8) Polysaccharides usually play the role of signal molecules in cellular activities; peptides in the framework of the organizational structure also have the role of signal conduction, mediating physiological activities, but the study of natural polymers in this regard is still relatively shallow; polysaccharides and peptides in the field of bioengineering are often simply used as a material scaffold, but their effective physiological and immunological mechanisms are not explored yet.

References

1 Zou, W.J., Chen, Y.X., Zhang, X.C. et al. (2018). Cytocompatible chitosan based multi-network hydrogels with antimicrobial, cell anti-adhesive and mechanical properties. *Carbohydr. Polym.* 202: 246–257.
2 Lei, Z.Y., Zhu, W.C., Zhang, X.C. et al. (2021). Bio-inspired ionic skin for theranostics. *Adv. Funct. Mater.* 31: 2008020.

3 Ruiz-Esparza, G.U., Wang, X.C., Zhang, X.C. et al. (2021). Nanoengineered shear-thinning hydrogel barrier for preventing postoperative abdominal adhesions. *Nano Micro Lett.* 13: 212.

4 Ouyang, J., Ji, X.Y., Zhang, X.C. et al. (2020). In situ sprayed NIR-responsive, analgesic black phosphorus-based gel for diabetic ulcer treatment. *Proc. Natl. Acad. Sci. U.S.A.* 117: 28667–28677.

5 Chen, S.G., Zhang, S.B., Galluzzi, M. et al. (2019). Insight into multifunctional polyester fabrics finished by one-step eco-friendly strategy. *Chem. Eng. J.* 358: 634–642.

6 Vergaro, V., Scarlino, F., Bellomo, C. et al. (2011). Drug-loaded polyelectrolyte microcapsules for sustained targeting of cancer cells. *Adv. Drug Deliv. Rev.* 63: 847–863.

7 Jin, L., Zhang, X.C., Li, Z.R. et al. (2018). Three-dimensional nanofibrous microenvironment designed for the regulation of mesenchymal stem cells. *Appl. Nanosci.* 8: 1915–1924.

8 Li, J.G., Li, Z.R., Chu, D.D. et al. (2019). Fabrication and biocompatibility of core–shell structured magnetic fibrous scaffold. *J. Biomed. Nanotechnol.* 15: 500–506.

9 Li, Z.R., Chu, D.D., Chen, G.X. et al. (2019). Biocompatible and biodegradable 3D double-network fibrous scaffold for excellent cell growth. *J. Biomed. Nanotechnol.* 15: 2209–2215.

10 Li, Z.R., Zhang, X.C., Ouyang, J. et al. (2021). Ca^{2+}-supplying black phosphorus-based scaffolds fabricated with microfluidic technology for osteogenesis. *Bioact. Mater.* 6: 4053–4064.

11 Luo, S.L., Wu, S.Y., Xu, J.M. et al. (2020). Osteogenic differentiation of BMSCs on MoS_2 composite nanofibers with different cell seeding densities. *Appl. Nanosci.* 10: 3703–3716.

12 Li, Z., Zhang, X., Guo, Z. et al. (2021). Nature-derived bionanomaterials for sustained release of 5-fluorouracil to inhibit subconjunctival fibrosis. *Mater. Today Adv.* 11: 100150.

13 Yang, Z., Gao, D., Guo, X.Q. et al. (2020). Fighting immune cold and reprogramming immunosuppressive tumor microenvironment with red blood cell membrane-camouflaged nanobullets. *ACS Nano* 14: 17442–17457.

14 Kong, N., Zhang, H.J., Feng, C. et al. (2021). Arsenene-mediated multiple independently targeted reactive oxygen species burst for cancer therapy. *Nat. Commun.* 12: 4777.

15 Ji, X.Y., Ge, L.L., Liu, C. et al. (2021). Capturing functional two-dimensional nanosheets from sandwich-structure vermiculite for cancer theranostics. *Nat. Commun.* 12: 1124.

16 Gao, D., Chen, T., Chen, S.J. et al. (2021). Targeting hypoxic tumors with hybrid nanobullets for oxygen-independent synergistic photothermal and thermodynamic therapy. *Nano Micro Lett.* 13: 99.

17 Liu, C., Sun, S., Feng, Q. et al. (2021). Arsenene nanodots with selective killing effects and their low-dose combination with β-Elemene for cancer therapy. *Adv. Mater.* 33: e2102054.

18 Yang, J.L., Zhang, X.C., Liu, C. et al. (2021). Biologically modified nanoparticles as theranostic bionanomaterials. *Prog. Mater. Sci.* 118: 100768.

19 Gao, D., Guo, X., Zhang, X. et al. (2020). Multifunctional phototheranostic nanomedicine for cancer imaging and treatment. *Mater. Today Bio* 5: 100035.

20 Li, J., Song, S., Meng, J.S. et al. (2021). 2D MOF periodontitis photodynamic ion therapy. *J. Am. Chem. Soc.* 143: 15427–15439.

Declaration of Competing Interest

The authors declare that they have no known competing financial interests or personal relationships that could have appeared to influence the work reported in this paper.

Natural Polymers for Biomedical Applications, First Edition. Wenguo Cui and Lei Xiang.

Nomenclature

AAm	acrylamide
ASCs	adipose stem cells
BADSCs	brown fat-derived stem cells
BC	bacterial cellulose nanofibers
CA	cellulose acetate
CG	catechin gallate
CNF	cellulose nanofibers
CSMA	methacrylated chondroitin sulfate
EC	epicatechin
ECG	epicatechin gallate
ECM	extracellular matrix
EGC	epigallocatechin
EGCG	epigallocatechin gallate
ESA	engineered stromal cell-derived factor analog
G	α-L-guluronic acid
GBM	glioblastoma multiforme
GC	gallocatechin
GCG	gallocatechin gallate
GelMA	methacrylic acid gelatin
GGA	antioxidant gallic acid conjugated gelatin
GH	gelatin-hydroxyphenylpropene acid
GSH	glutathione
GST	glutathione transferase
HaCaTs	human keratinocytes
HAMA	methacrylic hyaluronic acid
HA-SpyTag	SpyTag-containing elastin-like polypeptide modified hyaluronic acid
HDFs	human dermal fibroblasts
HiPSCs	human-induced pluripotent stem cell
HIV	human immunodeficiency virus
HMSCs	human marrow mesenchymal stem cells
HUVECs	human umbilical vein endothelial cells
LBL	layer-by-layer
M	β-D-mannuronic acid

Natural Polymers for Biomedical Applications, First Edition. Wenguo Cui and Lei Xiang.

MMPs	matrix metalloproteinases
MSCs	mesenchymal stromal cells
NIPAM	*N*-Isopropylacrylamide
PACG	cleavable poly(*N*-acryl-2-glycine)
PAM	polyacrylamide
PCL	poly-3-caprolactone
PDMAAm	poly(*N*,*N*-dimethylacrylamide)
PECDA	acryloyl chloride-poly(ε-caprolactone)-poly(ethylene glycol)-poly(ε-caprolactone)-acryloyl chloride
PEG	polyethylene glycol
PEGDA	poly(ethylene glycol) diacrylate
PEGTA	polyethylene glycol-tetraacrylate
PLGA	polylactic acid-glycolic acid copolymer
PRF	platelet-rich fibrin
PTFE	polytetrafluoroethylene
QK	a mimic of the VEGF receptor-binding region and full sequence: KLTWQELYQLKYKGI
RGD	arginine–glycine–aspartic acid
ROS	reactive oxygen species
VEGF	vascular endothelial growth factor

Index

Natural Polymers for Biomedical Applications, First Edition. Wenguo Cui and Lei Xiang.